To Chiraporn, Nairn and Nuan

COST AND FINANCIAL ACCOUNTING IN FORESTRY

A Practical Manual

by

KEITH OPENSHAW

B.Sc. (For) Hons; M.A. (Econ) Hons

*Associate Professor
Division of Forestry, University of Dar es Salaam
and
International Forest Science Consultancy*

PERGAMON PRESS

OXFORD · NEW YORK · TORONTO · SYDNEY · PARIS · FRANKFURT

U.K.	Pergamon Press Ltd., Headington Hill Hall, Oxford OX3 0BW, England
U.S.A.	Pergamon Press Inc., Maxwell House, Fairview Park, Elmsford, New York 10523, U.S.A.
CANADA	Pergamon of Canada, Suite 104, 150 Consumers Road, Willowdale, Ontario M2J 1P9, Canada
AUSTRALIA	Pergamon Press (Aust.) Pty. Ltd., P.O. Box 544, Potts Point, N.S.W. 2011, Australia
FRANCE	Pergamon Press SARL, 24 rue des Ecoles, 75240 Paris, Cedex 05, France
FEDERAL REPUBLIC OF GERMANY	Pergamon Press GmbH, 6242 Kronberg-Taunus, Pferdstrasse 1, Federal Republic of Germany

Copyright © 1980 Keith Openshaw

All Rights Reserved. No part of this publication may be reproduced, stored in a retrieval system or transmitted in any form or by any means: electronic, electrostatic, magnetic tape, mechanical, photocopying, recording or otherwise, without permission in writing from the publisher.

First edition 1980

British Library Cataloguing in Publication Data

Openshaw, Keith
Cost and financial accounting in forestry.
1. Forests and forestry—Accounting 2. Cost accounting
I. Title
657'.863   HF5686.F62   78-41183

ISBN 0–08–021456–8 Hardcover
ISBN 0–08–021455–X Flexicover

Printed in Great Britain by Fakenham Press Limited, Fakenham, Norfolk

PERGAMON INTERNATIONAL LIBRARY
of Science, Technology, Engineering and Social Studies
*The 1000-volume original paperback library in aid of education,
industrial training and the enjoyment of leisure*
Publisher: Robert Maxwell, M.C.

COST AND FINANCIAL ACCOUNTING IN FORESTRY

A Practical Manual

THE PERGAMON TEXTBOOK INSPECTION COPY SERVICE

An inspection copy of any book published in the Pergamon International Library will gladly be sent to academic staff without obligation for their consideration for course adoption or recommendation. Copies may be retained for a period of 60 days from receipt and returned if not suitable. When a particular title is adopted or recommended for adoption for class use and the recommendation results in a sale of 12 or more copies, the inspection copy may be retained with our compliments. The Publishers will be pleased to receive suggestions for revised editions and new titles to be published in this important International Library.

CONTENTS

List of Figures, Tables, Examples and Formulae   ix

Introduction   xiii

Part I   The Costing of Forest Operations   1

Chapter	1	Standard account heads 3
	2	Primary records 10
	3	Depreciation 18
	4	Preliminary analysis of primary records 31
	5	Direct unit costs 39
	6	Overhead costs 54
	7	Usefulness of costs 70

Part II   The Financial Account   77

Chapter	8	Income and expenditure (trading) account 79
	9	Capital valuation 92
	10	Profit and loss account and balance-sheet 121

Part III   The Financial Yield   125

Chapter	11	Discounted expenditure 127
	12	Discounted income 137
	13	Financial yield 143
	14	Financial yield of the enterprise 154
	15	Net discount revenue 159
	16	Increasing profit 165

Appendix I   Standard account headings   170

Appendix II   Capital valuation comparison between actual and potential value methods   182

Index   185

LIST OF FIGURES, TABLES, EXAMPLES AND FORMULAE

Figures

Introduction	Fig. 1a.	Flow diagram of a woodland accounting system	xiv
	Fig. 1b.	Use of accounts	xv
Part II			
Chap. 9	Fig. 2.	Capital valuation forest plantation	96
	Fig. 3.	Capital valuation of a normal forest using various assumptions	114
Part III			
Chap. 12	Fig. 4.	Discounted-income curve	142
Chap. 13	Fig. 5.	Discounted-income and expenditure curves	144
	Fig. 6.	Discounted-income and expenditure curves (infinity)	147
	Fig. 7.	Measuring financial yield	148
Chap. 15	Fig. 8.	Net discount revenue of two equal-aged projects at varying rates of interest	162
	Fig. 9a.	Net discount revenue of two projects—single rotation	163
	Fig. 9b.	Net discount of two projects—infinite rotations	163
Chap. 16	Fig. 10.	Relationship between spacing and volume production	167
Appendix 1I	Fig. 11.	Capital valuation comparison	184

Tables

Part I

Chap. 1	Table 1.	Extract from Standard Head form	5
Chap. 2	Table 2a.	Time-sheet	13
	Table 2b.	Completed time-sheet	14
Chap. 4	Table 3.	Completed wage and operation analysis sheet	32
	Table 4.	Wage summary sheet	35
	Table 5.	Operation analysis sheet	36
Chap. 5	Table 6.	Compartment costing schedule	42
	Table 7a.	Compartment record card (establishment)	48
	Table 7b.	Completed record card (establishment)	50
	Table 8.	Operation costing schedule—planting	52
Chap. 6	Table 9a.	Overheads allocation form	56
	Table 9b.	Completed overheads allocation form	58
	Table 10.	Supervisor's time-sheets	63
	Table 11.	Overhead costing schedule	64

Part II

Chap. 8	Table 12a.	Income and expenditure account—Expenditure	80
	Table 13a.	Machine and running cost account 1978	84
	Table 13b.	Machine capital account 1978	85
	Table 14.	Roads account 1978	87
	Table 12b.	Income and expenditure account—Income	89
Chap. 9	Table 15a.	Capital valuation per unit area using historic costs	100
	Table 15b.	Capital valuation per unit area using current costs	106
	Table 16.	Capital valuation—total area	110
Chap. 10	Table 17.	Profit and loss account	121
	Table 18.	Balance-sheet	122

Part III

Chap. 11	Table 19.	Capital expenditure	130
	Table 20.	Multipliers used to calculate discounted expenditure	131

LIST OF FIGURES, ETC. xi

Chap. 11 (*cont.*)	Table 21.	Discounted expenditure (single rotation)	132
	Table 22.	Effect of rotation length and discount rate on discounted expenditure	133
	Table 23.	Multiplier used to discount back from infinity	134
	Table 24.	Discounted expenditure (infinity)	136
Chap. 12	Table 25.	Price/size relationship	137
	Table 26.	Standing value of crop	138
	Table 27.	Volume and money yield of crop	139
	Table 28.	Multipliers used for discounting income	140
	Table 29.	Discounted income (single rotation)	141
Chap. 13	Table 30.	Discounted income (infinity)	146
	Table 31.	Discounted income comparison, single rotation and infinity	146
	Table 32a.	Costs, revenues and financial yield on a 23-year-old spruce plantation	149
	Table 32b.	Inflation rate per year over the lifetime of the plantation	150
	Table 32c.	Financial yield on the spruce plantation after inflation	151
Chap. 14	Table 33a.	Financial return, Arbor enterprise	155
	Table 33b.	Extract from a compartment record card	157
Chap. 15	Table 34.	Net discount revenue	160
Chap. 16	Table 35.	Cost of planting at different spacings	168
Appendix II	Table 36.	Capital valuation comparison	183

Examples

Part I			
Chap. 3	Ex. 1.	Straight-line method of depreciation	20
	Ex. 2.	Declining-value method of depreciation	21
	Ex. 3.	Fixing depreciation rate for the declining-value method	22
	Ex. 4.	Sum of the digits method of depreciation	23
	Ex. 5.	Production methods of depreciation	23
	Ex. 6.	Production methods based on the declining-value method	24
	Ex. 7.	Annuity method of depreciation	25

LIST OF FIGURES, ETC.

Chap. 3 (*cont.*)	Ex. 8.	Methods of calculating interest payments	27
	Ex. 9.	Methods of calculating the purchase price of a new machine taking inflation into account	28

Part II
Chap. 9	Ex. 10.	The determination of the limits bounding the appropriate rate of interest	104
	Ex. 11a.	Costs and revenues calculation	113
	Ex. 11b.	Potential capital value calculation	113
	Ex. 12a.	Costs and revenues calculations	117
	Ex. 12b.	Expectation capital value calculation	118

Part III
Chap. 11	Ex. 13.	Calculation of infinity multiplier	135
Chap. 13	Ex. 14.	Determination of financial yield percentage	144

Formulae

Part I
Chap. 3	Form. 1.	Depreciation rate, declining-value method	22
	Form. 2.	Annuity factor	25

Part II
Chap. 9	Form. 3.	Financial yield rate of interest	112
	Form. 4.	Expectation value	116
	Form. 5.	Capitalisation value	119

Part III
Chap. 11	Form. 6a.	Compound interest formula	130
	Form. 6b.	Discount interest formula	130
	Form. 7.	Discount value of a constant yearly sum	131
	Form. 8.	Infinity formula	134
Chap. 13	Form. 9.	Financial yield determination by interpolation	143

INTRODUCTION

More and more attention is being paid to the management of forests, whether natural or plantation. Governments and individuals are questioning not only investment in forestry, but spending between various forest projects or management practices. The expected rate of return for any new project is demanded by governments or forest enterprises before schemes are approved and the benefits of previous investments are having to be quantified.

In order that forest practice may be critically assessed and compared with alternative investments, it is necessary not only to undertake sound silviculture, but to have a thorough knowledge of costs and prices. Unless detailed information is recorded and analysed, the manager will find it difficult to practise his skills in the most advantageous manner. Without accurate costs, forward planning and budgeting may be meaningless and if expected returns are not calculated, commercial forestry enters the realms of faith. Input/output and cost/benefit analyses cannot be tackled, nor can cost reduction or budget programming be undertaken if costs are not specific. Therefore, it is important for the forest manager to have a reliable and simple system of cost and financial accounting.

This manual attempts to describe and illustrate methods of cost and financial accounting. Financial accounting gives a precise periodic view of the whole venture. It presents a balance sheet stating assets and liabilities and shows whether the enterprise as a complete unit is making a profit or loss at a definite moment in time. Cost accounting, on the other hand, provides information about the costs of individual operations or products. It probes into details, distinguishes the profitable from the unprofitable lines and is able to guide the manager towards profit maximisation. In cost accounting it may be difficult to be precise for, as a rule, overhead costs—the managers', supervisors' and office costs—cannot be allocated accurately. Nevertheless, for comparative purposes direct costs can be a reasonable substitute for total costs and in marginal analysis overhead costs are superfluous, so these two components will be dealt with separately.

In this book emphasis is laid on costs because this is where management can have most impact and because generally speaking the forest manager cannot affect the price structure. However, he should have a thorough knowledge of prices in order to obtain the most favourable market price for his product. Once individual operation costs are known, the woodland manager, with the help of timber prices and management (yield) tables, can anticipate the financial yield from individual woodlands and forests or the total enterprise. He can also look for ways of increasing efficiency and reducing costs. It can be seen, therefore, that a good system of cost and financial accounting is an important tool in forest management.

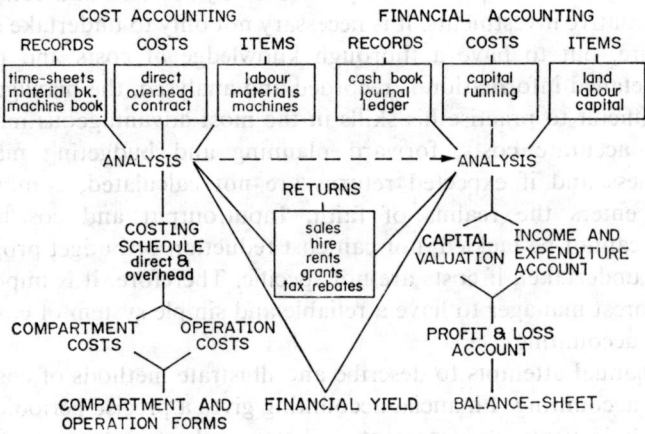

Fig. 1a. Flow diagram of a woodland accounting system.

Figure 1a is a flow sheet of the various components of woodland accounts. It gives a schematic picture of the costing of forestry operations, showing how the various costs—direct and overhead—can be extracted from different books and records. Time-sheets, contractors' accounts, machine (log) books and the material (forester's day) book are the records required to obtain direct operation costs, whereas the income and expenditure account could be obtained from the cash book and journals or ledger, or built up from the direct and overhead costs. Likewise the returns are recorded and extracted in a similar way.

These costs can then be analysed and unit costs both direct and overhead for labour, materials and machines calculated and recorded by compartment and operation. Also a profit and loss account plus balance-sheet may be drawn up from the valuation of the growing stock and income/expenditure records. Finally knowledge of cost and price enables the manager to work out Financial yield forecast.

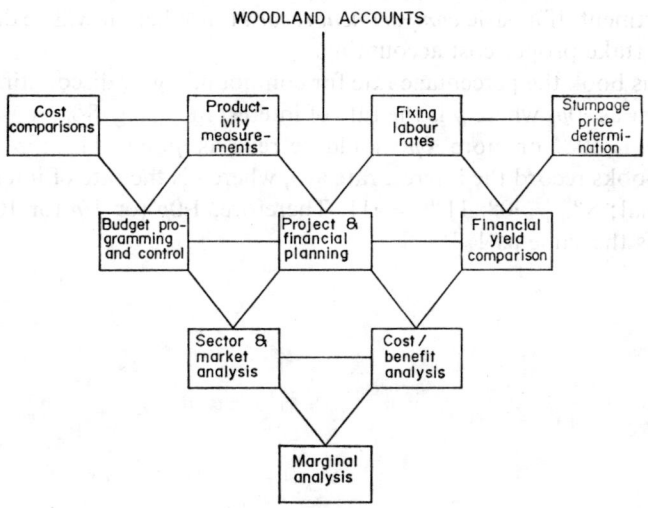

Fig. 1b. Use of accounts.

Figure 1b shows how the information from the analysis may be used for cost comparisons, fixing labour rates and measuring productivity, budget programming and control, various analyses, stumpage appraisal, financial yield comparisons and finally financial planning. Accurate accounting, therefore, plays a major part in scientific management and planning.

Any accounting system is difficult to describe briefly, and different countries lay stress on different aspects, due to varying taxation laws and perhaps climatic variations. However, the principles remain the same and, once they have been mastered and a routine established, the system is very easy to apply. When writing this book the author decided that it should be illustrated as much as possible with examples—examples that were related to one another so that they may be easier to follow. Hence U.K. money values and practices have been used, but the readers may substitute particular examples relevant to their countries once basic principles have been understood. The reader may find it easier to follow the links and get a feel of the system if the detailed examples are passed on a first reading, and then follow the examples the next time round.

It is assumed that an accurate record and map is kept of the forest which lists the area, planting year, species and soil type, etc., of each compartment. If a basic *compartment record* is not kept it will be difficult to undertake proper cost accounting.

In this book the percentage rate for compounding or discounting, etc. is shown as $\cdot 0p$, where p is the rate of interest in %, e.g. 8%; $\cdot 0p = \cdot 08$. When p is 10% or more $\cdot 0p$ should be read as $\cdot p$, e.g. 11%; $\cdot p = \cdot 11$. Some books record the interest rate as i, where i is the rate of interest as a decimal; $8\% = \cdot 08$, $11\% = \cdot 11$. Therefore, $1 \cdot 0p$ (or $1 \cdot p$ for 10% or more) is the same as $1 + i$.

Part I
THE COSTING OF FOREST OPERATIONS

Chapter 1

STANDARD ACCOUNT HEADS

Introduction

When keeping basic records it is logical to divide the different items into various sections or accounts according to the type of transaction. This is done to minimise error, for the sake of order, to facilitate checking, and for ease of application and analysis. These sections are commonly known as *Standard or Account Heads*.

In financial accounting broad categories such as land, labour and capital for the nursery, woodlands and sawmill are sufficient divisions under which to record the various items because this type of accounting is concerned with the enterprise as a whole. It does not require a detailed breakdown by operation: its purpose is to present a balance-sheet showing assets, liabilities and overall profitability.

While this is an adequate system for financial accounting it cannot serve as a basis for cost accounting because the account heads—land, labour and capital—are not related to physical outputs such as the area planted or the volume felled. Cost accounting should provide information about the cost of individual operations or products. Therefore, these operations have first to be identified and then classified.

In forestry many operations are not precise and therefore a strict comparison between the same kind of operation may not be valid. For example, planting costs depend on the age of plants, method of planting, type of ground, whether the roots are exposed or in a tube, weather conditions and even species: to compare planting on ploughed ground using 2-year-old seedlings with notch planting on steep slopes with 4-year-old transplants without being aware of the different methods and conditions may lead to erroneous conclusions. It is for this reason that it is important to have a system of classification—*the Standard Head*—which embraces every kind of forestry operation and transaction.

The system of standard heads outlined in this book groups related operations under general headings such as establishment and tending for besides wanting costs of individual operations the manager will require the overall cost of establishment, tending, harvesting, etc. Each individual operation is given a standard head and a number; if subdivisions are necessary these too are identified. Such a system holds the key to the keeping of accurate and, in the end, time-saving records.

Purpose of Standard Heads

When people are filling in time-sheets, they often describe their work in an irregular and individual manner. Thus weeding may be described as "paring", brashing as "pruning", barking as "peeling" and so on. If such various descriptions of the same kind of work are separately entered in sectional accounts or in costings, the results tend to be confused and inaccurate. It is, consequently, necessary to classify and standardise the operations which are required in forestry and it is convenient to give these operations distinctive numbers.

The advantage of issuing standardised headings in a cost account system—apart from their brevity—is that they help to make sure that those who record the information and those who make the subsequent analysis are thinking of the same thing. The introduction of electric calculators and computers has necessitated the adoption of standard heads, and if a single uniform system were adopted for a whole country, standard programmes could be prepared which would be of national use.

Table 1 is part of a list of standard heads: the full list is given in Appendix I. There is nothing sacrosanct about this particular list but it has been prepared with care and may be accepted as a basis. The various operations are classified under such broad headings as establishment, tending, etc., and the numbers are so arranged as to be easy to remember. Thus, the sixties all refer to the nursery, the seventies to sawmill and yard, and so on. On every forest enterprise there will be special operations, usually of a local character, which are not included in the list; the blank numbers can be used for such operations or they can be indicated by placing a letter or an additional integer after one of the numbers.

STANDARD ACCOUNT HEADS

TABLE 1. *An Extract from the Standard Account Heads (Appendix I)*

EXPENDITURE

0–9 *Establishment (up to 5 years in the U.K.)*   *Units to be used when*
    (May be only 1 year in the Tropics)   *costing*

0 Fencing	(a)	Plain stock (Sheep, cattle,	Area and distance
	(b)	Rylock stock goats, etc.)	
	(c)	Rabbit and other burrowing animals	
	(a & c)	Rabbit and stock	
	(d)	Roe deer (and other leaping	
	(e)	Red deer animals)	
1 Draining	(a)	Main drains	Area and distance—
	(b)	Side drains	indicate if draining by
	(c)	Upkeep and maintenance as preparation for planting only	hand (i) or machine (ii)
2 Ploughing	(a)	Single furrow	Area
	(b)	Single furrow with tine	
	(c)	Single mould-board	
	(d)	Double mould-board	
	(e)	Discing	
3 Clearing	(a)	Scrub	Area
	(b)	Lop and top	
	(c)	Burning	
	(d)	Taugya/Shamba system	
4 Manuring	(a)	Hand	Area
	(b)	Mechanical	
	(c)	Aerial	
5 Other preparatory treatments			
	(a)	Turfing	Area and per turf
	(b)	Preparation for natural regeneration	
	(c)	Elephant trenches, etc.	
6 Artificial and natural regeneration (including screefing)			
	(a)	Planting	Area and number
	(b)	Direct sowing	(per 1000 trees)
	(c)	Natural regeneration	Area
7 Weeding (up to 5 years)			Area
8 Beating-up (replacement of dead trees) (up to 5 years) (includes replanting—100% B.U.)			Area and number (per 1000 trees)
9 Local or miscellaneous establishment operations			

TABLE 1 (cont.)

10–19 Tending (from 6th year onwards in the U.K.)	Units to be used when costing
10 Early cleaning: until formation of canopy—removal of all adverse growth	Area
11 Late cleaning: after formation of canopy—removal of woody-growth	Area
12 Beating-up (belated)	Area and number (per 1000 trees)
13 Enrichment planting (interplanting)	Area and number (per 1000 trees)
14 Underplanting	Area and number (per 1000 trees)
15 Manuring (belated)	Area
16 Rack-cutting (line cutting)	Distance
17 Brashing (a) 100% (b) 50% (c) 33%, etc.	Area
18 Pruning—high pruning by species	Area and number
19	

Generally, forestry operations are very variable, depending, for example, on the type of land to be planted or the markets for the thinnings and fellings. It may be impractical to specify in detail the precise operation without having too many standard heads. However, the more details that are recorded the more it is possible to compare and contrast similar operations and standard heads can facilitate the recording of this information.

If an enterprise has several types of fences, such as deer round the perimeter and stock dividing the fields from woods, different letters can be given to the different types of fences: for example, 0c—rabbit fencing, 0e—red deer fencing. One would expect these different types of fences to have different costs per unit length. Similarly, other major operations can be divided and suggestions as to the various subdivisions are given in the table. However, the forest manager should first concentrate on those operations where expenditure is greatest for it is generally in these areas where the most savings can be made.

Definitions

It is necessary to define precisely the meaning and scope of various operations. In this manual establishment of a plantation covers all the operations from fencing to beating up (the replacement of dead plants) —Table 1. The time span covered by this phase varies from country to country and species to species. In the Tropics it may be only 1 year whereas in Northern Europe 5 years may be the general time period. However, there are a few operations which one would normally associate with establishment that may be carried on beyond the anticipated time span, for example weeding and beating up. A distinction is made between these same operations by giving them different standard heads. Thus, in the United Kingdom, beating up carried out before the sixth year is given the standard head 8, whereas after the fifth year it is treated as a tending operation and has the standard head number of 12. Similarly, the removal of weeds during establishment is known as weeding and has the standard head number of 7, whereas the standard head numbers are 10 and 11 during the tending phase and are known respectively as Early and Late cleaning. The former is the removal of all adverse growth until the formation of the canopy, whereas the latter is the removal of woody growth after canopy formation.

Miscellaneous items will occur under every section. It is proposed that the blank numbers or the letter M should be used for these whether they occur in the woods, nurseries or elsewhere.

Harvesting and Conversion

Of all the work in the woods the most difficult part to cost is harvesting and conversion. This work is of great variety and it is often difficult to separate the time devoted to individual operations or products. This is particularly so when thinning young plantations. The usual procedure in this case is for the men to fell and trim thinnings and extract them to a road or ride; on the ride side the poles may be cut into pitprops, pulpwood, stakes, building poles, etc. In such an operation it is impossible to work out the exact cost of preparing an individual product such as pitprops. It can, however, be estimated by dividing the various costs by the total weight or volume of the separate products; this will give an approximate cost of felling and extraction, cross-cutting and delivery.

Similar confusion may arise if oak or wattle bark is being ripped for tanneries. It is important to know whether bark-ripping is profitable, but the fellers will be felling and ripping at the same time so that it is impossible to distinguish the particular costs of ripping. The best method in such a case as this is to estimate the cost of felling from the fellings in other woods and to deduct this cost.

In the standard heads for harvesting and conversion, many items such as bark, cordwood, stakes are not operations but products. It is intended that as far as possible all the costs of producing a product should be included under the appropriate head. These costs will include many operations such as cross-cutting, barking, extraction, etc., and these can be distinguished by placing a letter after the numbers of the standard head. Thus 35 refers to pitprops, 35(i) is cross-cutting them, 35(ii) barking them, and so on. This system of numbering can be used as far as is required for distinguishing particular parts of operations.

Record All Cost Components

When analysing the costs of the various operations, it is important to make sure that the entire operation is recorded and that the cost can be separated into its different components. It is futile to compare hand weeding with weeding by a machine or chemical weeding if the machine running costs or the chemical cost are not available. And again, if only the total cost of weeding by a machine or with chemicals was available, the increase in labour productivity, if any, could not be determined. *The method of performing the operation should be known—hand, machine, spray, etc.—and recorded.* The cost of the various components of an operation should be extracted from the wages analysis sheet, the vehicle log analysis book and the forester's day book. This would be straightforward if the appropriate standard head be placed after each entry. For example, Tractor, Standard Head 1b(ii), 8 hours C.3—a tractor used for (side) draining in Compartment 3.

When any contract work is carried out, this too should be recorded by standard head and compartment. The contractor should be asked to divide the costs into the various components; for example, fencing should be divided into labour, materials and material transport costs. However, it must be remembered that contract costs contain an element of contractor's overheads and profits. Therefore, any division of costs will not be a perfectly factual breakdown of direct charges.

Division of Standard Heads

In Appendix I the standard heads are divided into two main divisions: *Expenditure* and *Income*. The Expenditure division is further subdivided into *direct* and *overhead costs*. A direct cost as the name implies can be attributed directly to a specific operation. Overhead (or indirect) costs, on the other hand, are usually concerned with the management of the enterprise as a whole. Most headings are obviously either a direct or an overhead cost, but there are some where the division is not clear. Marking and measuring may be carried out by supervisory staff or the working head forester. If done by the former, the operation may be treated as an overhead item, whereas if carried out by the latter it may be regarded as a direct cost. For the sake of uniformity, all marking and measuring items are treated as overheads in this book and like all other overhead costs are depicted by CAPITAL LETTERS.

Similarly, if the employer pays compulsory weekly National Insurance, Health and Pension Premiums for each workman, these could be regarded as an overhead or a direct cost. However, this book treats these items as being *direct costs* and it is assumed that they are included in the costing of the various operations. For example, the direct cost of planting is so much per unit area and this cost includes the employer's share of National Insurance, etc.

The standard head list is rather long but any one group of workmen will be concerned with only small sections of this list. *It is essential that they are issued with and know the standard heads of their particular operations.* The use of standard heads will be fully illustrated throughout this manual.

Chapter 2

PRIMARY RECORDS

Introduction

The keeping of good but simple records is the key to easy accounting be it cost or financial. The list of standard heads given in Appendix I is purposely thorough in order to minimise error. Primary records may be filled in by many people and the chances of errors occurring will depend, in the final analysis, on the system being fully understood and the basic rules followed.

There are three principal primary records forms—forms which record both quantitatively and qualitatively the movement of labour, machines and materials. These are the time-sheets/job cards, log (machine) books and material record books.

In order that accurate analysis be undertaken, each primary record form should state for every job where the work has taken place and if materials or machines have been used. Generally the compartment or the area planted in a specific year is the most convenient measure in forestry and usually all costs, past, present and future, should refer to this unit.

It is necessary to record accurately the time spent on each operation, remembering that the larger the time unit chosen the more likelihood there is of error. The day or half-day may be the most convenient unit but more than one operation could take place in the period and its length may not be uniform: the morning half-day may be longer than the afternoon half-day; Saturday working day could be shorter than other week days. Therefore, the hour is recommended as the unit of time.

Time-sheets, log books and material record books do not have to be elaborate but it is advisable that every operation is distinctly recorded by compartment, and the movement of machines and materials should be documented separately. An example is set out in the following pages showing how a simple time-sheet may be filled in. Also the type of information needed in the machine and material record books is stated.

Time-sheets

Good primary records are the key to accurate costings. They need not be elaborate but should record the movements of labour, materials and machines, including contract work, on an enterprise. A *time-sheet* is a document that simply records the daily activities of the work force. Any enterprise should keep attendance records and little extra effort is involved in recording the work force's daily tasks. Ideally each worker should fill in his own time-sheet, but the forester, supervisor or clerk could perform this essential job.

With time-sheets, adequately and reliably filled in, it is quite practicable, years after, to ascertain the cost of individual operations on each specific area of an enterprise, even when book-keeping is of the simplest. Without these primary records this kind of information cannot be supplied by the most elaborate accounting system.

When filling in a time-sheet there are certain basic principles which must be adhered to if fruitful analysis is to be achieved. *Costing must be made to a specific unit—the compartment or area planted in that year.* It is essential that the person filling in the time-sheet should specify the compartment in which he is working. It is no use a forest worker stating that he was planting for so many days, or even planting in spruce wood for so many days. He should put down on his time-sheet the standard head (S.H.) 6a—Planting in Cpt. 2 spruce wood—and then fill in the hours or days.

Time unit—hour:day

It is important that the time unit be accurately recorded and that it should refer to the operation in hand. The best time unit is the hour, but some enterprises use the day or divisions of the day. It must be borne in mind that there may be more hours in the morning half-day than in the afternoon half-day, and if an equal cost rate be applied to each "half" inaccuracies will occur. Also, Saturday morning, if worked, should not be counted as a whole day for costing purposes.

Distinguishing Direct and Overhead Costs

It is necessary for precise costings to record all activities and to exclude overhead items such as wet time, sickness or holidays, from direct costs. Wet time (S.H. 110) should be entered as such on the time-sheets and not disguised as contributing to the direct cost of some such operation as weeding (S.H. 7). It may be difficult and it is not necessary to record every slight stoppage due to rain, but stoppages of over an hour should be recorded for the sake of accuracy.

The cost of transporting workers (S.H. 113) to and from their place of work—sometimes a large item on enterprises with scattered woodlands—is generally regarded as an overhead cost. However, the workers' pay while travelling is usually included in the direct cost of the operation concerned.

Time-sheet Design

Time-sheets do not have to be of an elaborate nature; they can be prepared using a duplicating machine. A specimen form, Table 2 (a and b), shows a typical individual time-sheet for one or two weeks. Table 2a is blank whereas Table 2b has been made out by an imaginary workman and illustrates a number of points:
1. A separate line is taken for each individual operation per compartment. There are two plantings (S.H. 6a) during the period but in different compartments so each has a separate line. Planting in Cpt. 2 took place on two separate occasions but the time spent is all noted on the same line, but in different columns.
2. During any one day a number of operations may take place; these are recorded and as a check the total of hours worked per day is noted.
3. At the end of every time-sheet period the total of hours per operation, per compartment, is added and the sum of these totals should check with the hours worked during the period. If piecework is

PRIMARY RECORDS

TABLE 2a. *Specimen Time-sheet*

..........................Forest enterprise

Time-sheet for one/two week(s) ending..................19..

Name........................ Work No........... Week No....

1st week						2nd week					Description of work	S.H.	Hours, days or P.W.	Where working
M	T	W	Th	F	S	M	T	W	Th	F	S			
											Total hours/days			
											Total piecework (P.W.)			
											Total overtime hours			

Worker's signature .... Checked by............
S.H. = Standard Head

Note

The wage and operation analysis sheet can be printed on the reverse side of the time-sheet. If properly designed it can be folded and the information transferred without constantly turning the time-sheet over (see Table 3).

employed, whether the payment be per 100 plants for planting or per unit length for fencing or draining, the units completed should be recorded and totalled.
4. If any overtime is worked, this should be recorded separately. Casual work and bonus payments should also be tabulated.
5. The kind of operation carried out should be indicated with precision and by an adequate description of the work and by the appropriate standard head.

TABLE 2b. *Completed Specimen Time-sheet*

Name: **J. SMITH** ....Arbor........Forest enterprise

Time-sheet for one (two week(s)) ending...**1st April 1978**......

Work No. ...**F3**...... Week No. ...**26**......

	1st Week							2nd Week							Description of work	S.H.	Hours days or P.W.	Where working
	M	T	W	Th	F	S	S	M	T	W	Th	F	S					
	8	4													Fencing: finish stock & rabbit fence in Pine wood	Oa & c.	12	C2
		4													Sawmill: cutting firewood during wet weather	74	4	S/M
			6	4											P.G.: cutting scrub with power saw in Pine wood	3a	10	C5
			2												Wet time	110	1	—
				4+1	4					6					Planting: J.L. as firebreak Spruce block 1 hr. O.T.	6a	14+1	C10
					4										Sick (Hospital)	112	4	—
								8							Spring Holiday	111	8	—
									2	2					Beating up—P'76 area with S.P. in Pine wood	8	4	C1
									2						B.U.—P'73 with S.P. (fire damage) in Pine wood	12	2	C6
										4					Brashing—P'61 S.P. 100% in Pine wood	17a	4	C8
											P.W.	P.W.			Planting S.S. on ploughed ground	6a	2100	C2
											1100	1100			P.W. @ £5 per 1000 plants Spruce wood		plts.	
	8	8	8	8	8	—		8	8	8					Total Hours/Days	64	64	
											1100	1100			Total Piecework (P.W.)	2100	2100	
															Total overtime hours (O.T.)	1	1	

Worker's signature ......**J. Smith**......

Checked by ......**W. Brown**...... (Head Forester)

Key:
- S.H. = Standard Head.
- P.W. = Piecework.
- O.T. = Overtime.
- S/M = Sawmill.
- P.G. = Preparation of ground.
- B.U. = Beating up.
- P. = Planting year.
- plts. = Plants.
- S.P. = Scots pine *Pinus sylvestris*.
- S.S. = Sitka spruce *Picea sitchensis*.
- J.L. = Japanese larch *Larix japonica*.

Machine Record Book including Animals[1]

Records should be kept of all the enterprise machines, cars, tractors, lorries, timberjacks, scrub cutters, power saws, etc., in order to keep a strict account of running costs, depreciation and overheads. The cost charged to the various departments for the use of the machines should be sufficient to cover all these items. The machine record book (log book) is the usual form of account book in which to record the information—such information as:

		S.H.
1.	Purchase price of the machine and year of purchase	
2.	Value at the start and end of the year	78, 85
3.	Depreciation	
4.	Distance/time and purpose of use	77
5.	Fuel—volume and cost	
6.	Oil/lubricants—volume and cost	76, 81
7.	Electric power—quantity and cost	
8.	Animal feed (horse, elephant, etc.)—quantity and cost	82
9.	Insurance and licence—cost	84
10.	Repairs and maintenance—costs	75, 82
11.	Spare parts—costs	83
12.	Garaging—maintenance or hire costs	89

Each machine or each group of machines should be treated separately, the information being detailed under the appropriate head, ideally on a separate page in the log book. A time-sheet should be made out for each machine detailing its use, by time and place. Of course, it is usually impossible to work out a charge for a particular machine using current costs. Therefore, the previous year's costs should be used with suitable adjustments. If at the end of the year the machines have made a "book profit" this will be shown in the accounts and an adjustment will be made to the next year's rates. Similarly, with a "book loss". (Table 13.)

Other information that should be entered in the log book includes:

		S.H.
13.	Wages of driver or horseman and mate	80
14.	Travelling expenses	128
15.	Transport of materials	26, 27, 65, 77
16.	Transport of men	113
17.	Hire of vehicles and machines	86
18.	Carriage in or out	87

---

[1] An old but still useful article on this topic is to be found in *Unasylva*, Vol. 11, No. 3, 1957, by A. R. Patterson, entitled "An accounting system for mechanical and motorised equipment".

These items may be cross entries with other sections of the accounts or within the same section. For example, the distance entry (4) should record not only the mileage/kilometres but the purpose of the journey. If this is done, a cost can be placed on items 15 and 16 above.

Similarly for animals, their purchase price, training, equipment, maintenance, upkeep and job description should be recorded under separate headings using the appropriate standard head.

Material Record Book

The purchase of materials from outside the enterprise will be automatically recorded in the accounts book, and similarly, the sales of materials. The different items are shown in Standard Heads 90–109 (expenditure) and 200–239 (income). What should also be recorded are the "book sales" from one section of the forest enterprise to another or from the forest enterprise to some other department or vice versa. For example, the nursery transferring plants to the forest as well as selling to outside buyers; sales of timber to the enterprise out with the forest (e.g. the sawmill) as well as to the outside buyers. What is more, a realistic price should be placed on all products sold or transferred, irrespective of the purchaser whether for internal or external use. By not placing realistic figures on purchases and sales, one sector within the enterprise can be made to be very profitable or very unprofitable. A case in question is where all the timber from the woods is sold to the sawmill at a price just above the thinning and extraction cost. Naturally the sawmill should show a handsome profit whereas the woodlands never gets a chance to return other than a loss. Even if the sawmill and forest are combined and show a profit, nevertheless realistic costs and prices must be applied to the individual sections in order to show in which sections economies, etc., can be made. The whole idea of detailed costing is to highlight the economies of the different sections, plantations, compartments, species and methods. Unless realistic figures are attributed to these different factors, part of the truth will be masked.

A *Material Record Book* or *Forester's Day Book* should record all the materials obtained from the stores (S.H. 240–249) as well as material from the sawmill and nursery (S.H. 200–239). Separate pages should be used for fencing materials, plants, fertilisers, weed killers, etc. It will be noted that in the Standard Head income section it states that sales to outside buyers should be separated from sales to the forest sector and sales to the enterprise outside the forest sector. Therefore, it is proposed that Capitals A, B and C should prefix the Standard Heads, depending on where the sales are made. If the manager finds this confusing, separate numbers could be used for each standard head, depending on the buyer of the goods. Examples of entries from the forester's day book are shown in the miscellaneous section of Table 5.

Contractor's Costs

If work is done for the enterprise by outside labour, this contract work will be recorded in the accounts book. However, if more than labour is involved the contractor should be asked to itemise his bill into labour, materials and machines. These costs can then be added to the other costs when determining, for example, establishment costs. An example is given in Table 5. If the contractor is unable or unwilling to divide the costs into the various headings, then this should be attempted by the enterprise itself.

Chapter 3

DEPRECIATION

Introduction

The last chapter dealt with primary records and the type of information that should be recorded on these forms. Depreciation was mentioned in the section covering the machine record books and this chapter considers the reasons for and the methods of depreciation.

Reasons for Depreciation

Some assets such as labour and trees appreciate in value over time, the former through acquisition of knowledge and skills and the latter because of the dynamic nature of the forest growing stock. This capital appreciation has to be taken into account when valuing the assets of a forest enterprise. However, most assets decline in value through usage and due recognition must be given to this fact. Not all assets are depreciated: hand tools and consumable stores by tradition are written off as soon as they are purchased but equipment like vehicles and machines and other forms of capital expenditure such as buildings and roads are relatively expensive and generally have a lifetime in excess of one year, therefore they are depreciated. It would greatly distort yearly accounts if the total cost of such capital expenditure was debited against the income and expenditure account. It also means that the cost of the capital equipment is not evenly spread over its lifetime if the total cost is debited to a single year thus distorting costs and making cost comparisons meaningless. Depreciation is thus a method of dividing the investment over the useful lifetime of the capital equipment. This is perfectly legitimate and a sound practice. Depreciation is also a means of setting money aside to replace capital assets and it ensures that a realistic figure may be ascertained when calculating the total running costs of machines, etc.

The amount of depreciation depends on the *value* of the equipment, its useful *lifetime*, the *way* depreciation is calculated and the changes in the value of money.

The USEFUL LIFE of an asset is a relative term which is influenced by many factors. Theoretically a fixed asset should be employed as long as it continues to operate efficiently: it should come to an end when an alternative asset is more profitable to operate. Costs, obsolescence, production policy and other considerations will affect the decision. The *useful life* of an asset should be defined when the purchase is made, although advancement in, for example, machine design may in fact shorten the anticipated useful life.

Factors affecting Depreciation

There are many factors which may affect depreciation—time, intensity of use, obsolescence, etc. The main points may be summarised as follows:

1. Normal physical wear and tear—this mainly depends on intensity of use, "preventive" maintenance and routine checking. However, some wear depends on time and not use, perishable rubber for example.
2. Custom or usage—with some fixed assets there are customs which have been established on the rate of wear and tear normally expected each year, e.g. cars where second-hand prices may be published each month (also used for insurance purposes).
3. Abnormal occurrences—accidents, material defects and other contingencies (e.g. hairline cracks in boilers).
4. Technological changes and development. Machinery and material development may accelerate the obsolescence time.
5. Changes in production factors—labour rates rising relative to equipment means a shift to more capital intensive equipment.
6. Changes in the value of money and the interest rate. Inflation and/or changes in interest rate may prolong or shorten the useful life of equipment.

From a cost analysis viewpoint it is more convenient if the sum of the depreciation, running and repair costs are more or less constant over time thus giving a uniform cost from year to year. The methods of achieving this result are now discussed. It should be noted that the decline in the value of an asset tends to be greater in the earlier years and costs on repairs and maintenance increase with increasing age. In fact a time may come when these costs are more than the machine is worth.

Methods of Depreciation

There are various methods but all are concerned with spreading the original costs over the useful life of the asset. Whichever method is chosen it should be *employed consistently* (not changed to suit particular financial years), *administratively convenient* (easy to understand and employ), and *regarded as a legitimate cost*, not simply as an adjustment of profit.

For most of the methods a prerequisite is that the serviceable life in years of the asset is forecasted and the residual value (or scrap value) at the end of its useful life is known. The common methods of depreciation are as follows:

1. *Straight-line method* (proportional or equal instalment method)

    In this method the value of the asset is depreciated in equal amounts over its lifetime, Example 1.

    Example 1: *Straight-line method of depreciation*
    Asset cost £15,000; residual value (scrap value) estimated to be £1500; useful life estimated to be 5 years.
    Net value of asset = £15,000–£1500 = £13,500
    Depreciation per year = £13,500/5 = £2700

    Advantages
    Simple to understand and operate.
    Frequently used in practice.
    Obsolescence factor can be taken into account by reducing the time period.

    Disadvantages
    Fixed assets do not wear out at exactly the same rate over time.
    Repairs and maintenance higher in later years.
    The above two facts in combination will distort the cost calculations.

DEPRECIATION

However, average repair and maintenance cost for the 5-year period may be estimated; this average combined with the straight-line method depreciation rate will give a uniform cost from year to year. It is also convenient to use this method for assets that wear out over long time periods such as buildings and roads.

2. *Reducing-balance method* (declining-value method)

This method takes a percentage or a proportion of the balance that is remaining, Example 2. This procedure results in a diminishing amount being charged each year. Also if the correct percentage rate has been chosen we should finish up with a value at the end of the useful life approximately equal to the residual or scrap value. Therefore the residual value need not be estimated at the outset.

Example 2: *Declining-value method of depreciation*

Assets cost £15,000; useful life 5 years; depreciation rate estimated to be 37%.

	Value at the start of the year	Depreciation (rate 37%)	Value at the end of the year
	£	£	£
Year 1	15,000	5550	9450
2	9450	3496	5954
3	5954	2203	3751
4	3751	1388	2363
5	2363	874	1489 (scrap)

Total depreciation—£13,511

Advantages

As for the straight-line method plus:

Largest annual amount is charged (logically) in the first year—this means that the sum of depreciation and repair/maintenance bills may be more or less constant from year to year for the latter increases with increasing age.

Disadvantages

The rate of interest chosen will affect the accuracy.

It should be fixed only after considering all factors.

A standard percentage for all conditions may be misleading.

An asset is never written off completely (this may or may not be looked on as a disadvantage).

The question arises as how the interest rate is chosen so as to reflect the actual wear and tear of the asset each year. A mathematical formula can be employed to arrive at the appropriate percentage,[1] and written-down values for taxation purposes may be used as a guide.[2] Some rates that are applied in the U.K. are as follows: Tractors $28\frac{1}{8}$%, van, lorries, cars 25%, generating plant 20%. However, a *rule of thumb* is to take the straight-line depreciation rate and increase it to between $1\frac{1}{2}$ and 2 times its value, Example 3.

Example 3: *Fixing depreciation rate for the declining-value method*
Asset estimated life 5 years. By the straight-line method this will depreciate at one-fifth of the value (or 20%) per year. Therefore the declining value rate will be between $1\frac{1}{2}$ and 2 times this amount, that is between 30% and 40%.

Conclusion: The reducing-balance method is a logical attempt to depreciate fixed assets along the lines which occur in practice. Revenue often does tail off at the end of life of an asset.

In addition, applying discount cash-flow concepts future money is worth less than present-day money. There is the added advantage of recognising the growing cost of repairs and maintenance. The reducing-balance method has many factors in its favour *and is the one that is recommended* in this book.

[1]The mathematical formula to find the percentage rate of depreciation is given below, formula (1).
Formula 1: Depreciation rate for the declining-value method

$$100\left(1 - \sqrt[n]{\frac{S}{A}}\right) \quad (1)$$

where: S = the scrap value (at least 1 unit),
A = cost of the asset,
n = life expectancy (useful life).
However, if the life expectancy is short and the scrap value approaches zero then the depreciation rate is extremely high. This will lead to a very uneven machine cost from year to year, being relatively high in the first year and relatively low in the last year. It is therefore advisable to use the rule-of-thumb formulae with the rate twice the straight-line method (see above) and write off all the residual value in the last year. Alternatively the straight line or the sum of digits methods could be applied.
[2]These rates are average figures and the ones allowed for taxation purposes. It may be that the conditions, under which an enterprise's machines or vehicles work, are abnormal and therefore the allowable rates are not realistic. This is when internal and external accounts may differ (see p. 30).

3. *Sum of the digits method* (reducing proportion method)

The sum of the digits method is a method similar to the reducing-balance method, but the percentage charged each year decreases and the rate is dependent on the useful life, Example 4. If there is any residual (or scrap) value this would be deducted from the original cost before the calculations are carried out. The depreciation rate is fixed as follows: Knowing the useful life, the sum of the years is totalled. For example, an asset having a life expectancy of 5 years: Sum of years $= 5 + 4 + 3 + 2 + 1 = 15$. The first year's depreciation amounts to the life expectancy divided by the sum of the years and so on, e.g. 5/15 1st year, 4/15 2nd year, ... 1/15 5th year.

Example 4: *Sum of the digits method of depreciation*
Asset value £15,000; scrap value £1500; life expectancy 5 years.

Year	Original cost less scrap value (if any) £	Sum of digits depreciation rate Fraction	%	Depreciation £
1	13,500	5/15	33	4455
2	13,500	4/15	27	3645
3	13,500	3/15	20	2700
4	13,500	2/15	13	1755
5	13,500	1/15	7	945
Total 15		1	100	13,500

The above method has similar advantages and disadvantages to the reducing-balance method but it is slightly more difficult to calculate and the depreciation reduces at a constant value (in the example by 1/15 each year).

Also the scrap value has to be determined beforehand.

4. *Production methods*

With all the above methods the depreciation is not related to the use of the machine. Some methods try to overcome this problem by estimating the number of units produced and/or the number of production hours, Example 5.

Example 5: *Production methods of depreciation*
Asset value £15,000; scrap value £1500; estimated units of production 100,000 units or 8000 hours (in 5 years).

(a) Production unit depreciation:
$$\frac{\text{Net value of asset}}{\text{Production units}} = \frac{£13,500}{100,000 \text{ units}} = £0.135/\text{unit.}$$

(b) Production hour depreciation:
$$\frac{\text{Net value of asset}}{\text{Production hours}} = \frac{£13,500}{8000 \text{ hours}} = £1.688/\text{hr.}$$

With this method the determination of the production units/hours is a major problem. Only after running several machines for their lifetime will an average figure be known, then it is possible that technological changes will make the machines under examination obsolete.

Advantages

This method attempts to equate service and depreciation. It is much more realistic to equate depreciation with work done.

Disadvantages

As for the straight-line method, obsolescence factor more critical, difficult to forecast production.

(c) Production unit/hour based on the declining-value method.

In cost accounting it is necessary to know with some accuracy the cost per hour or per unit of output. It is far easier to estimate production per year (or time/distance per year) than total lifetime production. Such production, after an initial running-in period, usually declines over time but for any one year, knowing the age of the vehicle or machine and operating conditions, an average figure may be assessed from previous records, Example 6. Alternatively the actual production may be totalled at the end of the year and this figure used.

Example 6: *Production methods based on the declining-value method*

Asset 1 year old; value start of year £9450, estimated depreciation (declining-value method) £3496. From previous records estimated working hours during the second year = 1800 hours.

Production hour depreciation:
$$\frac{\text{Year's depreciation}}{\text{Year's production hours}} = \frac{£3496}{1800} = £1.942.$$

The average variable costs may be determined using either estimates or the actual figures and therefore a total running cost per unit of output can be given. However, when costing vehicles and machines one usually cannot wait until the year is complete, especially on con-

DEPRECIATION 25

tract work, and therefore accurate estimates are necessary. Production depreciation based on the declining-value depreciation is favoured because production also declines with age as well as variable cost increasing.

5. *Annuity method* (amortisation method)

The above methods do not take into consideration the rate of interest if money has to be borrowed, or if it is considered that the money used for capital equipment purchase should earn income. Whether this should be done is discussed later.

Interest can be considered separately in the above methods and added to the depreciation but there are methods which include the rate of interest in the calculation: the most common is the annuity method, Example 7. This method considers the original cost and interest on the *written-down* value of the asset. In effect the assumption made is that the purchase of a fixed asset is an investment on which interest is earned. The "investment" for the purpose of the method is the written-down value plus interest earned to date. A fixed rate of interest is determined beforehand and charged throughout the life of the asset. The annuity is calculated using formula (2).

Formula 2: Annuity factor:

$$\frac{V \times 0.0p(1.0p)^n}{(1.0p)^n - 1} \quad [3] \tag{2}$$

where: V = net value of the asset,
p = rate of interest in %,
n = life expectancy,
or $V \times$ annuity factor.

The annuity factor may be worked out or obtained from tables.

Example 7: *Annuity method of depreciation*

Asset £15,000; interest rate 8%; scrap value £1500; life expectancy 5 years.

With $p = 8\%$ and $n = 5$ years, the annuity factor = 0.25.
Annual depreciation = £13,500 × 0.25 = £3375.
Total depreciation = £3375 × 5 = £16,875.

This sum is £3375 more than the total depreciation by former

---

[3] May be written as $\dfrac{V \times i(1 + i)^n}{(1 + i)^n - 1}$ where i = rate of interest as a decimal; see page xvi.

methods and is accounted for by the interest at 8% on the written-down value of the asset over 5 years.

Advantages

Gives full recognition that a fixed asset is regarded as the purchase of a future flow of cash (an annuity).

Due recognition is given to the earning power of the fixed asset.

However, as mentioned previously, interest may be included in the other methods.

Disadvantages

An equal amount is depreciated each year and therefore it has the faults of the straight-line method. The rate of depreciation depends on the rate of interest adopted and it is difficult to vary it throughout the anticipated life of the asset. The interest is only charged on the outstanding depreciation value whereas logically it should be charged on the original loan for it is this value that was invested. Also with plant, vehicles and machinery the changes which take place such as sales, renewals and additions tend to make the annuity method difficult to apply.

Interest

The annuity method is based on artificial assumptions relating to the earning of interest; in most cases the interest as such never materialises but it shows up in the profit of the firm (if any). Assuming that interest will be earned is to anticipate a profit, which is contrary to accounting conventions. Capital equipment is purchased in order to earn money for the firm (just as labour and land, etc., is employed) and this will be reflected in the profit (or loss) at the end of the year.

There are other forms of capital investment, especially in forestry, and if interest is applied to this investment it should be applied to all investments. For example, when plantations are established an investment is being undertaken which will not be realised for many years; some of this investment is in the form of labour and materials as well as machines. Therefore if interest is to be charged on capital investment it should be charged on all capital investment not just capital equipment. It is wrong in principle to single out one factor of production and treat it separately from other factors.

Again it is only at the start of an enterprise or when an expansion is taking place that money may have been borrowed for capital equipment. At other times the money set aside as depreciation should cover the cost of purchasing replacement equipment, provided inflation has been taken into account.

When considering investment in forestry or any other enterprise, yield calculations should be undertaken to determine the likely returns on the proposed investment. Such calculations should exclude interest for the point of the exercise is to see if the returns are greater than a certain minimum or the next best alternative: logically these returns should be in excess of current interest rates. Obviously if interest is to be included as a cost then financial yield will be reduced. Therefore, for most purposes the *exclusion of interest* from depreciation costs is to be preferred. However, when comparing two different forms of capital inputs, for example hand saws versus power saws, the inclusion of interest may be necessary for in one case money may have to be borrowed to finance the purchase of capital equipment but its inclusion should not be a permanent feature. Nevertheless, interest payment on loans and capital repayments have to be included in the general accounts and provision must be made for their payment.

However certain countries, for example the Scandinavian countries, always include interest on capital equipment as a cost hence the reason for describing the annuity method. Interest payments may be built directly into the depreciation process without using the annuity method and this is illustrated below, Example 8. A short-cut method may be used when working out the average annual repayment for machinery equipment. This method takes 60% of the interest rate and applies it to the full loan instead of working out for each year the interest payment on the outstanding loan. (This is assuming that interest is only paid on the outstanding loan repayments and not on the original full amount.)

Example 8: *Methods of calculating interest payments*

Loan £15,000 borrowed at 10% from the bank. Repayment at the rate of £3750 per year for 4 years.

   (a) *Short-cut method:*

      Interest on original loan = 60% of 10% = 6%; 6% of £15,000 = £900 per year interest plus £3750 capital repayment.

(b) *Normal method:*

	Outstanding loan £	Interest repayment at 10% £	Capital payment (end of year) £
1st year	15,000	1500	3750
2nd year	11,250	1125	3750
3rd year	7500	750	3750
4th year	3750	375	3750
Total	—	3750	15,000
Average		938	3750

Note the 60% short-cut method is a rough guide only to average annual interest payments. If the interest rate on loans varies from year to year then the short-cut method is more approximate. In such cases it is better to use the normal method.

Inflation

The repurchase price of capital assets may be much greater than the money set aside by depreciation if inflation has been high during the lifetime of the asset. On average the purchase price of goods will double every 7 years with an average inflation rate of 10%. Therefore this factor may have to be built into the depreciation formula *when setting money aside* for repurchase. A rate of interest similar to the increase in price of the particular asset should be applied to the purchase price each year and added to depreciation or the average inflation rate could be compounded up for the period on the original purchase price. Alternatively the current purchase price of a new machine of the same model could be ascertained and the difference in purchase price between one year and the next could be recorded as the inflation addition. This is the value in the end column in the following example, Example 9, assuming the price in the previous column is the current purchase price.

Example 9: *Methods of calculating the purchase price of a new machine taking inflation into account*

Asset purchase price	£15,000
Life expectancy	5 years
Inflation rate per year	see Table
Depreciation method	reducing balance
Normal depreciation	£13,511
Normal scrap value	£1489

DEPRECIATION

Year	Purchase price of asset at the start of the year £	Inflation rate %	Value of asset if purchased at end of year £	Money set aside to account for inflation £
1	15,000	10	16,500	1500
2	16,500	12	18,480	1980
3	18,480	14	21,067	2587
4	21,067	12	23,595	2528
5	23,595	10	25,955	2360
	Av.	11.6	Total	10,955

Alternative:

£15,000 at 11.6% compound interest for 5 years = £25,966, an increase of £10,966.

It should be noted that the inflation rate is applied to the purchase price of the asset in that particular year, not the written-down value of the asset. This "inflation figure" has to be added to the normal depreciation figure of £13,511 plus the scrap value. However, it will be slightly in excess, for the original scrap value of £1489 will have grown to £2576 because of inflation. So the total money put aside for replacement will be as follows:

Depreciation	£13,511
Scrap value	£2576
Inflation	£10,955
Total	£27,042

Again it is to be expected that the money set aside for capital replacement will be invested and earn interest so the actual money available may be considerably more than the replacement value. But on the other hand, because of technological improvement a replacement machine may be more expensive as well as more efficient and it is not a bad thing to set aside money for improved equipment.

The Use of Depreciation in Cost Accounting

The forest manager requires the rate of depreciation on a particular vehicle or machine so that he can determine with some accuracy the total cost of a piece of equipment per unit of output. He can then compare different types of machines, say, two different makes of power saws; different methods of operations, timber extraction by animal tractor or timber jack; calculate whether it is worth while to purchase or hire, say, a tractor. Again the manager should be interested in cost reduction and without a knowledge of all costs this may be difficult to achieve. Depreciation is sometimes considered to be a fixed cost but it is only tied to the year by convention and for convenience. The amount a piece of capital equipment depreciates will depend very much on its use and how it is used. A tractor used for 2000 hours per year for extraction under rough terrain conditions will not last as long as a similar one working for 1800 hours per year for transporting logs to a sawmill on well-graded gently sloping roads. Different useful life times will have to be allocated to the varying conditions to account for the differences. The taxation authorities may have a fixed rate for tractor depreciation, therefore in the financial accounts, which are presented to outside authorities such as governments and shareholders, this rate will have to be used for both these tractors. However, the cost accountant is interested in real costs so the depreciation rate will be related to actual conditions and it may differ substantially from the taxation rate. This rate will be used for internal purpose and in such a case internal and external accounts will differ. Again inflation accounting may or may not be permitted by the taxation authority and the external accounts will have to be governed by the rules. However, it still should be used internally to arrive at a realistic cost for capital equipment. After all, all other costs more or less increase at the same rate as inflation.

Because depreciation per unit of output will vary from year to year this cost element is important for marginal analysis not only to discover the optimum output but also to find out if and when a machine should be retired. It can be seen, therefore, that the determination of accurate depreciation rates is most important in accounting.

Chapter 4

PRELIMINARY ANALYSIS OF PRIMARY RECORDS

Introduction

In the analysis of primary records certain procedures should be followed and a routine established. It is necessary for any organisation to go through the time-sheets or some similar record forms in order to compile the weekly/monthly pay packets of the labour force and a little additional work is all that is required to undertake costings by operation. Once the wage rates are known they can be used to calculate the direct cost of individual operations by a simple analysis of the time-sheets, for these sheets record the man hours worked by operation and compartment. Also the machine and material books record the items used in individual compartments during the period in question so that these items too can be costed by operation.

It may be useful for the forest manager to make a record of the man hours per operation as well when the analysis of the records is being undertaken. Such information is necessary for manpower planning and comparison of operations from year to year (and between countries) when wage rates are not static or comparable.

The wage rate may be composed of more than one part depending on established practice in particular countries. Every country will have a flat rate bargained for by the individual workers, the union or fixed by a wage council or government. In addition the employer may by law or agreement contribute towards accident, life and health insurance plus a pension fund for the employee. These additions may be a fixed sum or could be in proportion to earnings. Irrespective of which system is used such contributions, if made, should be added to gross wage of the individual—that is the wage before tax and other deductions—

because they are a direct cost and as such vary with the number of workers.

It is this figure—gross earnings plus employer's additions—that should be used when calculating the hourly/daily rate for each member of the work force. There may be slight complications if overtime and/or piece work is undertaken, for the pension and insurance contributions will have to be added to the agreed rates but this should present little difficulty to the office staff.

TABLE 3. *Specimen*

..Arbor....Forest Enterprise

Wage Analysis Sheet

Worker..J. SMITH..... UNITS £

	Wage rate per			
	week	day	½ day	hour
Wage rate per 40 hr week	20.15	4.03	2.01	0.50
Employer's share of N.I.[1]	1.28	0.26	0.13	0.03
Employer's share of pension[2]	0.57	0.11	0.06	0.02
Total	22.00	4.40	2.20	0.55

Total wage bill	1st Week		2nd Week		Total
	day	P.W.	day	P.W.	
Gross Wage	20.15		12.09	10.50	42.74
Overtime	0.75		—	—	0.75
National Insurance	1.28		0.77	0.51	2.56
National Pension	0.57		0.34	0.33	1.24
Total	22.75		13.20	11.34	47.29

Piece work (P.W.) 2100 plants at £5 per 1000.
1 hour overtime (week 1) at £0.75 per hour.

[1]In the U.K. this was a fixed contribution per week.
[2]In the U.K. this was a variable contribution per week; it depends on the gross income.

Now both these rates are combined and variable depending on income. However, the example has been included to illustrate the principle.

PRELIMINARY ANALYSIS OF PRIMARY RECORDS 33

A schematic presentation of the analysis of primary records is shown below:

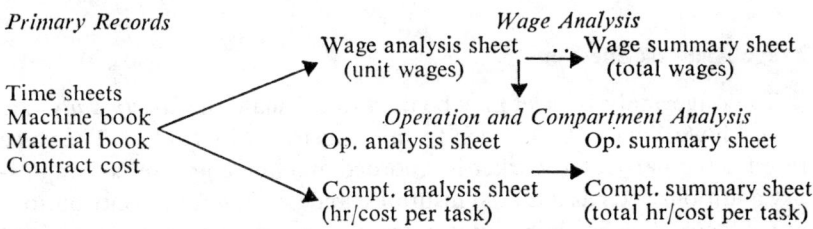

Primary Records

Time sheets
Machine book
Material book
Contract cost

Wage Analysis
Wage analysis sheet ... Wage summary sheet
(unit wages) (total wages)

Operation and Compartment Analysis
Op. analysis sheet   Op. summary sheet

Compt. analysis sheet   Compt. summary sheet
(hr/cost per task)   (total hr/cost per task)

Analysis Sheet

Operation Analysis Sheet

Work No. ..F3..   Week No. 26

Compt.	S.H.	Hours or P.W.	*Rate per hour	P. year	Labour cost
C1	8	4	0.55	76	2.20
C2	Oa & c	12	,,	78	6.60
C2	6a	P.W. 2100 p	—	78	11.34
C5	3a	10	0.55	78	5.50
C6	12	2	,,	73	1.10
C8	17a	4	,,	61	2.20
C10	6a	14+1	,,+(0.75)	78	8.45
S/mill	266 74	4	0.55	—	2.20
—	110	2	,,	—	1.10
—	111	8	,,	—	4.40
—	112	4	,,	—	2.20
		64	Total	→	47.29
		1 hr O.T.			
		2100 p			

*Includes employer's additions
S.H.   = Standard Head
P.W.   = Piece work
P. year = Year of planting

An example of how an analysis may take place in practice is illustrated in Tables 3–5.

Wage Analysis Sheet

The wage analysis sheet may be used to calculate the *hourly/daily* rate for each workman. A specimen form is illustrated in Table 3. The wage rate for the particular worker is recorded and any employer's compulsory additions such as national insurance and pension fund contributions are added, thus enabling unit rates to be calculated. As explained previously, these latter items are treated as direct costs because they are attributable to each workman.

Operation Analysis Sheets

Once the hourly/daily rate has been calculated, the labour cost for each operation can be determined. The operations on the analysis sheet should be transferred from Table 2 and systematically set down by compartments, direct costs being placed before overhead costs. The number of hours worked on each operation is recorded, together with the Standard Head, the wage rate and the plantation age (Table 3). Any piece work or overtime worked on the forest should be analysed in a similar manner and all costs referring to the same operation bracketed together (Table 3—C10 planting S.H. 6a, 1 hour overtime).

Summary Analysis Sheets

When the labour analysis has been completed the individual workers' costs are collected together on a master analysis sheet, totalled operation by operation and compartment by compartment (Tables 4 and 5).

Table 4 is the wage summary sheet for the forest labour force by operation and compartment. It shows the direct labour cost for a 2-week period for the various operations, as well as the total gross wage (including employer's additions) for the individual worker. These operation costs can be transferred to an operation analysis sheet, as is shown in Table 5 for weeks nos. 25 and 26. The analysis sheet has been divided into various sections according to category of operation.

Aston .... Forest Enterprise

TABLE 4. *Wage Summary Sheet by Operation*

One/two weeks ending... 1st April 1978 ...... Weeks No. 25/26

UNIT £

Compt.	C1	C2	C2	C2	C5	C6	C8	C8	C10	S/M	C/H	C/H	C/H	O/H	Total
S.H.	8	0a & c	1a(i)	6a	3a	12	11	17a	6a	266 (74)	110	111	112	120	
Forester F1	2.75	8.25	P.W. 600 m @ 0.07 45.40	12.98	6.88	1.37		2.75	6.60		1.37	5.50		8.25	56.70
F2				11.34	5.50	1.10				2.20	1.10	4.40			53.10
F3	2.20	6.60						2.20	8.45	2.20	1.10	4.40	2.20		47.29
etc.	↓								etc.	transfer to S/M account				→	etc.
Total	6.85	32.45	45.40	50.34	27.08	3.37	9.60	6.50	49.17	10.20	6.47	25.90	2.20	8.25	283.78

m = metre;   S/M = Sawmill;   O/H = Overhead;   P.W. = Piece work

Section I groups all woodland costs. It will be noted in the example that a contract ploughing cost is also shown under this section because it too refers to the woods. Section II is for the hire of forest labour to other departments, in this case the sawmill, and Section III for management.

The operation analysis sheet also contains details of the plants, machines, materials and transport use, together with their costs. These

TABLE 5. *Operation*

...A.T.Bot..... Forest Enterprise
one/two weeks ending ...1st April...1978

Compt.	Operation	S.H.	COSTS	
			Labour	Material
I.	*Woodlands*			
1	Beating up (B.U.)	8	6.85	7.70
2	Fencing	0a & c	32.45	49.22
2	Draining	1a(i)	45.40	
2	Ploughing	2d	<u>18.00</u>	
2	Planting	6a	50.34	111.60
5	Scrub clearing	3a	27.08	
6	B.U.	12	3.37	3.30
8	Late cleaning	11	9.60	
8	Brashing	17a	6.50	
10	Planting	6a	49.17	90.00
II.	*Sawmill/Nursery* (transfer item)			
	Firewood	266(74)	10.20	
III.	*Management and Overheads*			
	Wet time	110	6.47	
	Holidays	111	25.90	
	Sick	112	2.20	
2	Supervision—Drains	120	0.60	
2	Supervision—Planting	120	7.65	
	Total enterprise		283.78	261.82
	Total contract		<u>18.00</u>	

J.L. = Japanese larch; S.S. = Sitka spruce;
P.W. = Piece work; S/M = Sawmill;
Contract costs have been underlined.

PRELIMINARY ANALYSIS OF PRIMARY RECORDS   37

costs are determined in a similar way to the labour costs. Items bought in such as fencing wire are costed using the bills. Transferred materials—plants from the nursery and gates from the sawmill—should bear an element of profit in their costs. The charge should at least be comparable to the commercial prices, otherwise the justification of these sectors is doubtful. Materials produced by the enterprise itself, such as fencing

Analysis Summary Sheet

Week No. 25/26

Machine	Transport	Total	Miscellaneous Information
		14.55	700 SP 2+2
	0.62	82.29	Stock and rabbit 240 metres (m)
		45.40	Hand draining P.W. 600 m
213.00		231.00	Contract 10.5 ha @ £22 per ha
	3.00	164.94	9300 S.S. 1+1
12.00		39.08	3 power saws: 30 hrs @ £0.4 per hour
	0.10	6.77	300 SP 2+2 (P'69)
4.00		13.60	1 power saw: 10 hrs @ £0.4 per hour
		6.50	0.5 ha, 100% intensity
	2.43	141.60	7500 JL 1+1, 3 ha planted, fire break
		10.20	6 loads cut (18 tonnes) Charge to S/M Ac.
		6.47	
		25.90	
		2.20	
		0.60	Drain measuring C2
		7.65	Supervising P.W. planting C2
16.00	6.15	567.75	
213.00		231.00	

S.P. = Scots pine;   B.U. = Beating up (replacing dead plants);
2+2 = 4-year-old plants: 2 years as seedlings, 2 years as transplants.

posts, for internal use, should not include any profit but must be fully costed.

Machine and transport costs may be determined knowing or assessing depreciation, running, repair and maintenance costs. Transport costs have been separated from machine costs for they cover different cost components. The machine cost column records the costs of machines used to undertake forest operations such as power saws for scrub clearing and felling or tractors for ploughing and draining. The transport column records the cost of transporting materials and machines to the particular compartments—the plants from the nursery to the field and the plough from the garage to site. It may be difficult to separate transport costs for they may be included with the material or machine costs, or impossible to separate them from other cost components, for example, if the planting squad carry the plants with them when they go to the field. However, where possible the element of transport should be separated from other cost components to increase the accuracy of costing and to bring to the attention of the forester that it is a cost component.

On some enterprises, as is the case in this example, the nursery and sawmill labour forces may be independent of the forest labour force. Separate analysis sheets should be compiled for these two sections, similar to the ones shown above for the forest enterprise. The reason for including sections for nursery and sawmill costs (Table 5) is that the labour force between the various enterprises may be fluid. For example, during wet weather the "woods squad" may work in the sawmill or during busy times they may help the nursery squad out (and vice versa). It is important to allocate with accuracy the time spent between and within the various sections of the forest economy.

Chapter 5

DIRECT UNIT COSTS

Introduction

Cost analysis is the rearrangement of recorded cost data for the purpose of elucidating cost patterns and characteristics. Its purpose is to help management by instilling cost consciousness into all employees, comparing real costs with estimated costs, relating costs (inputs) to outputs (area planted, timber produced, etc.) and relating costs to benefits by examining the profitability of various methods of production.

The process of systematically recording information on time-sheets, etc., and undertaking a preliminary analysis has been dealt with in the previous chapters. Each year or when a specific operation is complete these periodic analysis sheets may be gathered together and unit costs/ man days worked out for individual operations or groups of operations.

In any cost analysis the unit has to be defined. In the establishment and tending stages of forestry it is usual to have the area as the standard unit but include other measures such as length and number. On the other hand, for production forestry volume and weight are the primary units with area and number as secondary measures.

The total operation cost may be difficult to obtain quickly and its accuracy is subject to the complete recording and allocation of indirect or overhead costs as well as direct costs. Overhead costs, which will be dealt with in the next chapter, not only include local supervision maintenance, office staff and equipment but some district, regional, headquarter and research costs if the project is part of a large organization. Therefore, direct costs—the cost of labour, machines and materials— may be used as a reasonable substitute for total costs for comparative purposes. Also direct costs, as such are required for the many aspects of management.

It is usual when working out unit costs to separate the various components, labour, materials, machines and transport. This enables comparisons to be made between the different inputs or factors of products and various outputs—the area planted or the volume felled. These costs are determined at specific time intervals and therefore comparisons could take place over time and patterns established.

There are a number of ways of undertaking the analysis of costs and a specific example is given in the following pages. A woodland costing schedule is shown on which direct unit costs are calculated. These costs are then transferred to compartment record cards and operation costing schedules so that a complete record of costs by operation and compartment is built up over time.

Woodland Costing Schedule

At the end of each forest year the operation summary sheets (Table 5) can be analysed and a compartment costing schedule compiled. Such a schedule lists all the operations carried out in the year and records the various costs. This is shown in Table 6.

It is convenient, but not necessary, that the financial year ends at the same time as the forest year (for the United Kingdom 30th September), but whereas the financial accounts have a strict closing date, the cost accounts are permitted a little more elasticity. For example, in the United Kingdom the growing season ends officially on the 30th September and in theory weeding in the plantation is over for the year. However, owing to a favourable summer for weed growth or shortage of labour, weedings may still be taking place after the official year has ended: in a case like this, for the purpose of cost accounting, the weeding costs will be allocated to the previous forest year. Normally, however, establishment operations which are repeated in another year or over-run into it will be charged to each year and when any one kind of establishment operation has been completed, its total cost can be ascertained by adding the year to year costs.

Table 6 analyses the direct costs of the enterprise and illustrates a number of points as follows:

   1. A reference number is given to each operation. At least one line should be allowed for each operation and, when clarity demands it,

a separate line for each sub-operation, such as one of those that constitute preparation of ground for planting (Op. Ref. Nos. 2, 3, 4—Table 6).
2. Should a plantation be weeded twice or more in the forest year, the costs are added together, the total being costed to the plantation area. A double weeding should be mentioned in the miscellaneous column (Op. Ref. No. 7). Weeding continued into October is included with that for the previous year; weeding started after October is a new operation. When an area that was planted over a period of years is weeded (or beaten up) it is usual to allocate the cost according to the areas planted in the respective years, unless information is available to enable a more precise distribution to be made.
3. When a date or an area can be given only in approximate figures, the degree of approximation should be indicated, e.g. "F.Y. '36 or '37", or "to the nearest half hectare". Wherever possible, area should be given to one decimal place. If an operation such as thinning covers more than one planting year, the years of planting should all be mentioned, the years with the smallest areas being placed in brackets (Op. Ref. Nos. 8–11).
4. The costs shown should be the costs actually ascertained for the area referred to. Any cost (such as the cost of a sub-operation) which is not itself an actual ascertained cost should not be entered unless differentiated by a note in the miscellaneous column (Op. Ref. No. 4).
5. When the work has been done by the enterprise, "labour costs" should consist only of wages plus the employer's additions. When the work has been done by a contractor the whole cost to the owner should be given, including contractor's charge for overheads. Contract costs are underlined to distinguish them from estate labour costs (Op. Ref. No. 3).
6. Definitions of the operations are given in the standard heads sections.
7. Some operations such as brashing and thinning may not cover the whole of a compartment in any one year. In a case like this only the area covered should be costed (Op. Ref. Nos. 8–10). Again, if fencing or ground preparation covers a greater area than the

42 COST AND FINANCIAL ACCOUNTING IN FORESTRY

TABLE 6. *Compartment*

...**Arbor**...Forest Enterprise

Op. Ref. No.	Cpt.	P year	S.H.	Costing Unit		Cost Components		
				Area (ha)	D. No. V. Wt.	Labour (L) £	L plus mat/mach £	L.m/m plus t'port* £
1	2	78	0a & c	10.5	1400 m	195.22	508.44	514.44
2	,,	,,	1a(i)	,,	2190 m	165.70		
3	,,	,,	2d	,,		18.00	231.00	
4	,,	,,	3a & c	,,		30.08	42.08	
5	,,	,,	6a	,,	31,200 p	168.88	543.28	553.28
6	,,	,,	7a	,,		26.25		
7	3	74	7a	5.0		75.00		
8	4	50(51)	23i	20.0	1760 m³	1348.00	1942.00	
9	,,	,,	23iii	,,	8340	674.00	871.00	
10	,,	,,	25	,,	stems	674.00	1268.00	
11	,,	,,	28	,,		54.00		
etc.								
17 etc.	10	78	0a & c	500.0	8760 m	1050.00	2900.00	2950.00
Enterprise cost 1978						6468.00	15,164.00	15205.00
Contract cost 1978						875.00	11,231.00	
36	7	55	42	50.0		16.00	34.00	

Key: Op. Ref. No. = Operation number
D. = Distance   mat = materials
V. = Volume   mach = machines
Wt. = Weight   S.H. = Standard head
No. = Number   p = pence (100p = £1)

Units: No. = per 1000 plants
D. = metres
Ha = hectares
V = cubic metres m³
stems = per tree

* Transport of materials may be calculated separately.

DIRECT UNIT COSTS

Costing Schedule (Direct)

Forest Year ending 30.9.78

Unit Costs								
Labour per			L+mat/mach per			L, m/m+transport		
Area (ha)	No(1000) V. (m³) Wt.	D (m) tree	Area (ha)	No(1000) V. (m³) Wt.	D (m) tree	Area (ha)	No(1000) V. Wt.	D (m) tree
£ 18.59		pence 14	£ 48.42		pence 36	£ 48.99		pence 37
15.78		7.5						
1.71			22.00					
2.86			4.01					
16.08	5.41		51.74	17.41		52.69	17.73	
2.50								
15.00								
67.40	0.77	16	97.10	1.10	23			
33.70	0.38	8	43.55	0.49	10			
33.70	0.38	8	63.40	0.72	15			
2.70								
2.10		12	5.80		33	5.90		33.5

Note: The remarks or miscellaneous column has been omitted due to lack of space. This column should contain such information as:
Op. Ref. 1: area fenced if more than the area planted, i.e. 10.7 ha.
Op. Ref. 3: contract work, labour cost estimated.
Op. Ref. 4: estimated burning cost £2.S.H.3c. 2.5 ha cleared S.H. 3a.
Op. Ref. 5: planting by piece work. Op. Ref. 6; only 2.5 ha weeded.
Op. Ref. 7: all area weeded twice. Op. Ref. 10; pine 0.21 m³/stem. Income £7/m³.
Op. Ref. 11: burning lop and top. Op. Ref. 36; fence repair, overhead cost.

planting in a particular year, the entire area is costed for the particular operation (Op. Ref. No. 17).
8. When scrub is being cleared for planting, some of it can occasionally be sold for firewood. It is the net cost of clearing that is used for costing purposes.
9. Expenditure on the following works and activities is regarded in this manual as overhead expenditure—to be costed to the total woodland area:

Repairs and Improvements: (a) Drains;[1]
(b) Fences;[1]
(c) Roads and Rides.
Protection: (a) from fire;
(b) from vermin (less sales);
(c) from insect and fungal pests.

The reason for treating these items as overhead costs is that they benefit the project as a whole and cannot be tied down to specific compartments.

When an "overhead" operation is entered on the costing sheet (Op. Ref. No. 36), it should be kept separate from the direct costs.
10. Two burning operations are shown—Op. Ref. Nos. 4 and 11. One is part of the process of preparing the ground for planting and the other is treated as part of the felling operation—burning lop and top. It is important firstly to classify the different operations, and once the operations have been classified, to stick to the classification.
11. The total estate costs are given at the foot of the page, together with the contractor's charges.

Unit area cost

Table 6 gives the total direct cost of labour, materials, machines and transport of materials for the forest year in question. Additional information, such as area covered and the length of fencing and draining, obtained from the time-sheets or forester's day book, is also indicated. This latter information is used in order to obtain the unit direct costs of the operation. The unit area cost is arrived at by dividing the various

[1]Except in preparation for planting.

DIRECT UNIT COSTS 45

labour and material costs, etc., by the *total area planted* in that year (i.e. 10.5 ha, Compt. 2). It will be noticed that in the example given the total area of Compartment 2 is 10.7 hectares. The unplanted 0.2 ha could be a pond or an outcrop of rock. Although 10.7 ha have been fenced, the costing is related to the 10.5 ha planted because this is the area of the commercial crop. Roads and rides are considered to be part of the commercial area.

There are exceptions to the above rule but these only occur when a compartment is planted over a number of years. An area that is to be planted over a number of years generally will have to be fenced prior to the first planting. In this case the fence costing is made to the total area to be planted. For example, 500 ha fenced in 1978; 250 ha planted 1978; 250 planted 1979; fencing area costed 1978—500 ha. Similarly, ploughing, draining and scrub clearing may cover an area greater than is to be planted in one year. In such cases the costing is related to the area actually covered, provided that this is the same as the total area to be planted over the period. What happens, for example, if an area of, say, 200 ha is to be planted over 2 years—100 ha per year—and only three-quarters of the area can be ploughed? Naturally, in order to obtain the cost per hectare ploughed, one would have to divide the ploughing cost by 150, but in order to obtain a unit cost as part of the ground preparation for the whole area, one would divide by 200 and this is the cost that would be shown in the Costing Schedule, with a mention of the other cost in the remarks column.

Table 6 also illustrates a number of other points. The costs shown in this table are direct cost only and include no overhead costs. Although the ploughing has been done on contract and this cost must include the contractor's overheads as well as his profit margin, nevertheless the whole amount constitutes a direct cost to the forest enterprise and is to be treated as such for the purpose of cost accounting.

Again, although the scrub cutting and weeding were restricted to a small part of the ground, these operations have been costed to the total area. If, however, the woodland manager wishes to fix piece-work rates, it is necessary for him to know a realistic unit area rate; hence the cost of the area actually weeded or cleared is given in remarks. This may be useful when comparing two types of soil—peat and sandy—or two types of treatment—ploughed and unploughed ground.

Other unit costs

Apart from "unit area" costs, other unit costs may be useful for comparative purposes and piece-work-rate determination. These include number (e.g. per 1000 plants) for planting and beating up and length (per metre) for fencing and draining, etc. These unit costs are shown in Table 6; the various costing units are given in the Standard Head List, Appendix I.

The fencing carried out in any particular area may not of itself form a ring round the whole plantation. It may complete the ring by adding two or three new sides to an existing fence. The unit length cost is the unit cost that should be used for comparison. However, the cost is still divided up according to area planted although the cost will be lower than if the whole perimeter had had to be fenced.

Compartment Record Card

Forestry, unlike most other businesses, is a long-term investment, and therefore one year's costs and returns will not give actual compartment profitability. Financial yield calculations can be made to estimate the return on a particular crop or area, but the only way to discover the actual profitability is to keep a record of costs and returns over the whole rotation of the compartment. Even so, costs and returns will have to be adjusted to account for inflation.

The easiest way to collate the compartment costs is to have a carding system on which the annual costs are recorded from the costing schedule (Table 6). An illustration of such a system is shown in Tables 7a and 7b for the "establishment" operations. Table 7a is a blank form showing how the information can be tabulated, whereas Table 7b gives an actual example of the area previously costed in Table 6.

A similar card can be used for the tending and harvesting operations so at the end of the rotation an "historic" and "inflation adjusted" account can be compiled.

In addition to the initial establishment cost, weeding and beating-up costs have been recorded so that the direct cost of establishment is shown. It must be noted that all the costs are related to the area planted in the specific year in question. If a greater area were fenced, then the fencing cost must be divided according to the area planted or to be planted in each year. The same method of apportionment is used with the other establishment operations.

Table 7a contains a column for overhead costs, and Table 7b gives these costs for the compartment in question. The way to determine these costs is described in Chapter 6.

Operation costs

The method of recording compartment costs has been described. Similarly operations or groups of operations such as initial establishment can be recorded in a tabular form and the results compared and contrasted. An example is shown in Table 8. This records the planting costs of Arbor enterprise for the 3-year period 1976–8, and similar tables could be compiled for other operations. Once a series of operation costs is known the manager could draw graphs of labour and total costs over time and if an allowance is made for inflation he could discover if productivity is increasing.

Table 8 tabulates three different labour methods of planting—day rate, piece work and contract—and therefore these methods can be compared, remembering that overhead costs must be included in the examination when contract labour is being compared with the local labour force. Again the cost of planting on differently prepared land (Op. Ref. Nos. 20a and 20b) and at various spacings (Op. Ref. Nos. 23 and 5) is shown to illustrate the different unit area costs. It must be remembered that planting on ploughed land may be cheaper both in labour and plant costs than other forms of planting because smaller plants can be used and the land has already been disturbed. However, it is false to compare only the planting costs; in a case like this establishment is the cost to be compared.

Average costs for individual operations may be determined from the operation cost sheets. These can then be used for budgeting purposes and to calculate the financial yield of plantations and forests. All in all it can be seen that accurate costings hold the key to sound management.

TABLE 7a. *Compartment Record Card (Establishment)*

...........Cmpt. Name........................ Area Planted ....

...........Yr. Planted Total Area ....

Money units ....

Year	Op. Ref. No.	S.H.	Operation	Cost		Remarks	O/H Cost	
				Total	Unit Area		Total	Unit Area
		0	Fencing					
		1 2 3 4 5	Draining Ploughing Clearing Manuring Other P.G. Treatment					
		1–5	GROUND PREPARATION TOTAL					
		6	Regeneration (Planting)					
		0–6	INITIAL ESTABLISHMENT					

DIRECT UNIT COSTS

	7	Weeding P+0					
	7	Weeding P+1					
	7	Weeding P+2					
	7	Weeding P+3					
	7	Weeding P+4					
		etc.					
	7	WEEDING TOTAL					
	8	Beating up P+1					
	8	Beating up P+2					
	8	Beating up P+3					
	8	Beating up P+4					
		etc.					
	8	BEATING UP TOTAL					
		Other preparations within period, e.g. belated manuring					
	0–9	ESTABLISHMENT TOTAL					
		OVERHEAD TOTAL					
		ESTABLISHMENT COST DIRECT & OVERHEAD					

NOTE Yr. = Year when the operation was carried out
Op. Ref. No. = The Operation Reference Number in that particular year
O/H = Overhead costs;   S.H. = Standard head;   P = Year of planting

TABLE 7b. Compartment Record Card (Establishment)

Cmpt. ...2... Name ...Arbor forest enterprise... Area Planted ...10.5 ha...
Yr. planted ...1978... Roadside Shelter Block Total Area ...10.7 ha...
Money units ...£...

Year	Op. Ref. No.	S.H.	Operation	Cost Total	Cost Unit Area ha	Remarks	O/H Cost Total	O/H Cost Unit Area ha
1978	3	0a & c	Fencing	514.44	48.99	Stock and rabbit	81.11	7.72
1978	4	1a(i)	Draining	165.70	15.78	2190 m P.W.	70.36	
	5	2d	Ploughing	231.00	22.00	Double mouldboard C	3.11	
	6	3a & c	Clearing	42.08	4.01		15.11	
		4a	Manuring	—	—		—	
		5	Other P.G.	—	—		—	
	1–5		P.G. TOTAL	438.78	41.79		88.58	8.44
1978	7	6a	Planting	553.28	52.69	31,200 plants S.S. 1+1	70.71	6.73
1978		0–6	I.E.	1506.50	143.48		240.40	22.89

DIRECT UNIT COSTS

Year			Operation				
1978	8	7a	Weed. P+0	26.25	2.50	2.5 ha only	13.41	1.29
1980	1	7	Weed. P+1	—			14.00	1.33
		7a	Weed. P+2	27.15	3.54	2.5 ha only	12.00	1.14
		7	Weed. P+3	—			14.00	1.33
		7	Weed. P+4	—			14.00	1.33
		7	WEED. TOTAL	63.40	5.04		67.41	6.42
1980	2	8	B.U. P+1	—		600 S.S. 2+1		
		8	B.U. P+2	12.22	1.16	¼ ha burnt	13.00	1.24
		8	B.U. P+3	—				
		8	B.U. P+4	—				
		8	B.U. TOTAL	12.22	1.16		13.00	1.24
			Other ops.					
1978–82	0–9		ESTABLISHMENT TOTAL	1582.12	150.68			
			O/H TOTAL	320.81	30.55		320.81	30.55
			ESTAB: COST (DIRECT & O/H)	1902.92	181.23			

Note: The per hectare overhead costs have been ascribed to weeding in years of no operations.

TABLE 8. *Operation*

......Ardot..........Forest Enterprise

Op. Ref. No.	Compt.	P. year	Costing Unit		Costs (direct)		
			Area (ha)	Plants (1000)	Labour (L)	L and plants(p)	L, p and transport
18	15	76	65.0	162.5	1218.75	3087.50	3117.50
23	15	77	65.0	162.5	877.50	2746.25	2776.25
5	2	78	10.5	31.2	168.88	543.28	553.28
20	10	78	250.0	625.0	3396.13	10896.13	10927.63
(20a	10	78	247.0	617.5	3346.98	10756.96	10787.96
(20b	10	78	3.0	7.5	47.17	139.17	139.67

Op. Ref. No. 20 has been divided into 20a and 20b.

Nursery and sawmill costing schedules

Nursery and sawmill costing schedules can be compiled in a similar way to that shown above. The big difference between these sectors and the forest enterprise is that short time spans are involved and so production costs can be determined quickly, and profitability of individual lines measured with certainty. However, just because large time spans are involved in forestry this does not detract from the importance of costings—it probably makes them more essential.

DIRECT UNIT COSTS

Costing Schedule—Planting

Units £

Unit Cost						O/H Costs	Remarks
Labour		L+plants		L, p+transport			
ha	1000 plants	ha	1000 plants	ha	1000 plants		
							p.p.ha
18.75	7.50	47.50	19.00	47.96	19.18	28.90	C. S.S. 1+1 2500
13.50	5.40	42.25	16.90	42.71	17.08	327.85	P.W. S.S. 1+1 2500
16.08	5.41	51.74	17.41	52.69	17.73	70.71	P.W. S.S. 1+1 2970
13.58	5.43	43.58	17.43	43.71	17.48	1470.80	P.W. S.S. J.L. 2500
13.55	5.42	43.55	17.42	43.66	17.47	1450.20)	S.S. 1+1
16.39	6.56	46.39	18.56	46.56	18.62	20.60)	J.L. Fire break

Key: C = Contract labour   S.S. = Sitka spruce
 P.W. = Piece work   J.L. = Japanese larch
 p.p.ha = plants per hectare
 1 + 1 = one-year seedling, one-year transplant

Chapter 6

OVERHEAD COSTS

Introduction

Overhead costs, sometimes known as fixed or indirect costs, unlike direct costs, cannot be allocated specifically to the unit of production be it the area of forest or the volume of timber removed. There are exceptions to this rule for by tradition some "direct-cost operations" which are undertaken by the supervisory staff are placed under the general category of overhead costs. For example, this book considers marking and measuring operations to be overhead items for they are usually undertaken by the forest officer. However, as will be explained, these costs should be distinguished from other overheads and placed in their correct category.

Overhead costs may be classified according to origin or the organisational level. The first classification divides the costs into similar groups such as labour additions, insurance, maintenance, office expenses, protection, rent, research, staff salaries, travelling, etc. The second specifies at what level the overhead costs are incurred—project, district, regional, headquarter.

Both classifications are useful for giving a general picture of where and at what level the overhead costs are incurred. They highlight where the money is being spent and should indicate to the manager where savings may be made.

However, these broad classifications are not particularly useful if an accurate allocation of overhead costs is desired. There are many suggestions as to how overhead costs should be allocated, for example on an area basis, in proportion to direct costs, according to man-days/ labour costs, relative to volume cut, or proportional to revenue. Different results will be obtained depending on which method is chosen and these results will not be consistent from year to year because overhead costs do not depend on any one of the factors listed above. Also when calculating financial yield different answers will result depending on the method of allocation of overheads. For these reasons it is recommended that the major factors which influence overheads should be picked out and the costs allocated accordingly. Some overhead costs are incurred to benefit or protect the whole project such as road construction, fire protection, insect control, boundary-fence repair. Clearly these costs should be allocated on an area basis and costed to each compartment *whether or not any activity has taken place in that particular area that year*. Other operations are closely related to direct labour costs such as labour additions (wet time, holidays) and part of the supervisory staff wages. Logically these costs should be divided according to direct labour charges (or man days) rather than total costs. Volume and revenue parameters are not good measures to use when allocating overheads for in some newly established projects production and hence revenue may not be forthcoming for several years. Finally there are those groups of overheads which are in fact direct costs—marking and measuring for felling, measuring establishment operations for assessing piece work. These costs are placed in the overhead category for they are generally carried out by supervisory staff but they can be specifically allocated to particular operations. In order to divide overheads into their respective categories—direct, proportional to labour costs and area—it is necessary for the supervisory and office staff to keep records or time-sheets of their activities for they may be involved in all three kinds of overhead activities.

An example of how to collect and analyse overheads according to the principles mentioned above is shown below. Once these overhead costs have been divided they may be added to the direct costs of each operation and posted on the compartment and operation record card thus enabling the total costs for any compartment or operation to be worked out for each year.

Overheads Allocation Form

Once the overhead costs relating to the woodlands have been ascertained, there are a number of ways of allocating them such as on an area basis or in proportion to direct costs. While it is true that some items must be costed on an area basis—maintaining roads, boundary fences, general protection against vermin, fire insurance, etc.—other overhead items are directly related to expenditure or income. The establishment and production phases in a plantation require far more supervision and administration than does the "early" growing stage. Therefore, *the woodland manager should himself keep a time-sheet* of his various activities as well as making his office staff do likewise. Only in this way will he surely discover the demands made on supervisory and office time by one operation as compared with another and know how to set about rationalising some of the operations and reducing their overhead costs. No attempt is made to show how the manager should set about making a time-sheet for himself and the office staff because sufficient illustrations have been given in the previous chapters. However, an overhead allocation form is shown in Tables 9a and 9b, Table 9b being a worked example for the Arbor enterprise. It must be pointed out that the scope of this Overheads Allocation Form is limited to the woodlands. Similar forms can be compiled for the nursery and the sawmill.

TABLE 9a. *Overheads Allocation—Woodlands*

Year ending ....

		Cost	S.H.
A.	**LABOUR ADDITIONS** (*including transport of workers*)		
	Wet time		110
	Holidays with pay		111
	Illness pay		112
	Transport of workers (to and from work)		113
	Pensions paid to former employees		114
	Employer's liability insurance		115
	Employer's contribution to Pension Fund or superannuation scheme		116
	Perquisites (e.g. free fuelwood)		117
	Workers' houses—cost less rent (if any)[a]		118
B.	**SUPERVISION**		
	Administrator's expenses		
	Salaries or proportion of salaries of:		
	Head Forester (if not a working Head Forester)	(20/21)	120
	Woodland Manager/Agent		121
	Office Staff		124
	National Insurance (employer's share), illness and holiday pay for above if charged separately		125
	Employer's liability, insurance and Pension Fund for above		126
	Annual cost of houses for Manager, less rent (if any)[a]		127

OVERHEAD COSTS

TABLE 9a (*cont.*)

	Cost S.H.
Travelling expenses for Manager, Head Forester, Owner	128
Value of Owner's time spent on supervision and administration (if any)	122
Director's and/or Consultant's fees	123
Perquisites for above	129

Office expenses
Rent and rates of premises	130
Light, heat and cleaning in office	131
Communication equipment (stationery, telephone, etc.), plus depreciation on office machinery	132
Books and technical magazines	133

Miscellaneous expenses
Audit and legal fees	134
Subscriptions to societies	135
Insurance premiums, on office, house, public liability, etc.	136
Bank charges, excluding interest on overdraft	137
First-aid kit, bad debts, etc.	138

C. *TOOLS*
New, repair and maintenance	88

D. *REPAIRS AND IMPROVEMENTS*
Construction of woodland boundary fence where no fences for current planting	40
Woodland boundary hedges	41
Maintenance of all woodland fences	42
Drains—upkeep (except as preparation for planting)	43
Road construction (depreciation, 1/40th for the U.K.)	44
Roads upkeep	45
Rides construction	46
Rides upkeep	47
Fire-dam construction	48

E. *PROTECTION*

Fire
Fire-guard and patrol expenses	50
Fire fighting	51
Fire-fighting equipment, preparation, maintenance and renewal	52/53
Fire insurance	54

Other
Vermin catcher's wages (or proportion) less sales	55
Subscription/payment to vermin eradication society/organisation	56
Control of insect and fungal pests	57
Stores, equipment, etc. (traps, etc.)	58

TABLE 9a (cont.)

F. LANDS AND TAXES, ETC.	Cost	S.H.
Rent		140
Land Tax		141
Product Tax		142
Interest on overdraft		143
TOTAL OVERHEAD COSTS		
TOTAL DIRECT COSTS		
GRAND TOTAL		
TOTAL OVERHEAD COSTS AS A PERCENTAGE OF DIRECT COSTS		

[a] Realistic rental value, e.g. for the United Kingdom, not less than the Gross Annual Value (G.A.V.) plus rates less rent (if any).

TABLE 9b. *Overheads Allocation—Woodlands*

..............Forest Enterprise   Year ending 30.9.78

	£
A. LABOUR ADDITIONS *(including transport of workers)*	
Wet time (Sawmill £210)	70
Holidays with pay	420
Illness pay	42
Transport of workers (to and from work)	Nil
Pensions paid to former employees	104
Employer's liability insurance	9
Employer's contributions to Pension Fund or superannuation scheme	Nil
Perquisites	273
Workers' houses—cost less rent (if any)[a]	420
	1338
B. SUPERVISION	
Administrator's expenses	
Salaries or proportion of salaries of:	
Head Forester (part)	325
Woodland Manager/Agent (21%)	1047
Office Staff (23%)	480
National Insurance (employer's share), illness and holiday pay for above if charged separately	—
Employer's liability insurance and Pension Fund for above (21%)	35
Annual cost of houses for Manager, less rent (if any)[a] (21%)	63
Travelling expenses for Manager, Head Forester, Owner (21%)	36
Value of Owner's time spent on supervision and administration (if any)	Nil
Director's and/or Consultant's fees	Nil
Perquisites for above (21%)	35
	2021

OVERHEAD COSTS

TABLE 9b (*cont.*)

	£
Office expenses	
Rent and rates of premises (21%)	10
Light, heat and cleaning in office (21%)	21
Communication equipment (stationery, telephone, etc.) plus depreciation on office machinery (calculator, typewriter) (21%)	32
Books and technical magazines	2
	65
Miscellaneous expenses	
Audit and legal fees (21%)	9
Subscriptions to societies (excluding vermin clearance)	10
Insurance premiums on office, house, public liability, etc.	5
Bank charges, excluding interest on overdraft	2
First-aid kit, bad debts, etc.	3
	29

C. **TOOLS**
New, repair and maintenance — **30**

D. **REPAIRS AND IMPROVEMENTS**

	£
Construction of woodland boundary fence where no fences for current planting	Nil
Woodland boundary hedges	Nil
Maintenance of all woodland fences (Lab. 16.00: Mat. 18.00)	34
Drains—upkeep (except as preparation for planting)	Nil
Road construction (depreciation 1/40th)	Nil
Roads upkeep ⎫	
Rides construction ⎬	
Rides upkeep ⎬	Nil
Fire-dam construction ⎭	
	34

E. **PROTECTION**

Fire	£
Fireguard and patrol expenses	50
Fire-fighting	Nil
Fire-fighting equipment, preparation, maintenance and renewal	5
Fire insurance	100
	155
Other	
Vermin catcher's wages (or proportion) less sales	Nil
Subs./payment to vermin eradication society/organisation	24
Control of insect and fungal pests	Nil
Stores, equipment, etc. (traps, etc.)	4
	28

TABLE 9b (*cont.*)

		£
F. *LAND, TAXES, ETC.*		
Rent	1360 ha @ £1.25	1700
Stipend (Tithe redemption)		4
Tax,[b] Schedule B	360 ha @ £0.10	36
Schedule D		Nil
Interest on overdraft		Nil
		1740
	TOTAL OVERHEAD COSTS	£5440
	TOTAL DIRECT COSTS	£26436
	GRAND TOTAL	£31876
	TOTAL OVERHEAD COSTS AS A PERCENTAGE OF DIRECT COSTS	21%

[a] Realistic rental value (not less than G.A.V.) plus rates, less rent (if any).
[b] Schedule B and Schedule D taxation is part of the taxation system in the U.K. Schedule D is for commercial businesses and B for non-commercial. Usually for tax purposes young (loss making) woodlands are under D and mature woodlands under B.

Table 9a has been divided into a number of groups which are self-explanatory—Labour Additions, Supervision, Tools, etc.—and the appropriate standard heads have been placed alongside each entry. In most cases the allocation of overhead expenditure can be made with a fair degree of precision but there are certain items which as a rule have simply to be estimated as closely as possible. For example, the overhead form refers only to the woodlands. The Forest Manager might have to look after a sawmill and nursery as well as some non-forestry jobs. How is the one to allocate the office expenditure, the manager's house rent, the audit and legal fees, etc.? The way described in this manual is to allocate the various expenses in proportion to time spent by the manager in his different roles. For example, Table 10b shows that out of a total yearly salary of £5000 the manager spent £1047 or 21% of his time administering the woodlands. Therefore, this figure of 21% has been used in order to arrive at the sum to be charged to the Woodlands Section of the various overhead items concerned (see Table 9b).

In a strict sense improvements, as an item of capital expenditure, should not appear in total in an annual account because the effect of the improvement usually lasts for many years. Therefore, they should be depreciated over a number of years. Depreciation is just a simple method of spreading capital expenditure over the anticipated life of the investment. But it may be that data for previous capital expenditures which would have to be taken into such a reckoning are unlikely to be found in the office files. If it could be assumed that capital expenditures on woodlands were fairly regular items annually, or in a short period, the distortion in the accounting caused by the inclusion of capital items would be less serious. This is usually the case, but there is one major exception and that is road construction. This is a costly operation and most unlikely to be continuous. If a road is laid down just prior to first thinning then, apart from a little maintenance, it should last till the end of the rotation and without major repairs for at least the lifetime of another crop. Therefore, for the purposes of overhead allocation, roads are depreciated at a standard rate (in the United Kingdom 1/40th of their cost). The road costs should therefore go into the capital account. Likewise, if the manager wishes to be strictly accurate, bearing in mind what has been said previously, other improvement costs should be entered in the capital account and depreciated in a similar manner, the depreciation cost appearing in the overhead schedule.

In Section F under Land and Taxes, provision is made for rent. In most cases the land may not be rented but be in the hands of the State or the owner. Nevertheless, a realistic rent per unit area should be charged to the enterprise. When fixing this rent the capitalisation value of the purchase price may be used or the rates charged for similar land in alternate use, e.g. sheep farming, could be adopted.[1] Section F also makes provision for taxes paid, both land and product, as well as interest on overdraft.

Taxes present a difficult problem when writing a book of this kind because each country has different laws which have evolved around the particular institutions in that country. However, most countries recognise that forestry, being a long-term investment with relatively long periods without income from specific crops, cannot be treated as a normal business. Therefore, special taxation systems are usually in operation for forestry enterprises. These may take the form of delayed payments, dual taxation systems which will, for example, allow rebates

---

[1] The concept of land as a resource in forestry and the different notions of rent are well summarised in *Economic Problems in Tropical Forestry* by A. J. Leslie, F.A.O., Rome, 1971.

on young immature woods while at the same time impose a nominal tax on the mature timber or even tax exemption. Some countries which have a small area under forest or want to encourage the development of forest industries may give grants, interest-free loans, or tax incentives, etc.

The manager or the student should find out the tax laws in their particular country and use them to the maximum benefit of the enterprise.

Table 9 gives a worked example of overhead allocation for the woodland. A breakdown of the manager's time-sheet, head forester's supervision time and office staff wages sheet, Section B—Supervision, could be as shown in Table 10.

Allocation of overheads

It is clear from Table 10a, b and c that some of the overhead costs can be charged directly to the operations and compartments concerned. Thus in Table 10—X, the overheads costed to the compartments and operations fall into this category, the *Direct Charge* category.

Again, it is logical that "labour additions" (Table 9) be allocated *proportionally* to labour costs because these overheads were incurred as a result of the operations carried out in that year. Other overhead costs that could be allocated proportionally to labour costs are: the Head Forester's discussions with the manager (£88.00); Time-sheet analysis (£236.00); Meetings (£140.00); Supervision (£400.00); Wage analysis (£390.00) (Table 10—Y). Therefore, in the example shown the total overhead costs to be allocated in the above way—the proportional category, amount to £2592 (£1388 of labour additions (Table 9b) plus the above sums). With accurate accounting the Woodland Manager will become expert at allocating the different overhead items to the various sections.

The total *direct labour costs* in the example shown is £6468 (Table 6). Therefore, the overhead cost to add on to each direct labour charge amounts to

$$\frac{2592}{6468} \times 100 = 40\%.$$

It would perhaps be more accurate, where the direct labour costs include such payment methods as day rate, piece-work rate, and bonus

OVERHEAD COSTS

TABLE 10. *Extract from Supervisor's Time-sheets*

(a) *Head Forester's Supervision Time by Operation (assumed to be mainly a working head forester)*

		£		S.H.
C.2	Drain measuring (P.W.)	0.85		1a
C.4	Measuring timber	70.00		23, 25
C.10	Counting turfs (P.W.)	1.75		5
C.10	Planting supervision (P.W.)	23.40		6a
C.16	Marking	37.00 ⎤		
	Measuring	23.00 ⎟		
C.19	Marking	23.00 ⎟		
	Measuring	9.00 ⎬		22, 36b
C.21	Marking	12.00 ⎟		25
	Measuring	14.00 ⎟		
C.25	Marking	16.00 ⎟		
	Measuring	7.00 ⎦	237.00X	
	Discussion meetings with Manager		88.00Y	
	TOTAL		£325.00	

(b) *Manager's Time-sheet—Analysis*

		£		S.H.
C.4	Measuring timber	105.00		23, 25
C.16	Marking	23.00 ⎤		
	Measuring	18.00 ⎟		
C.19	Marking	14.00 ⎬		22, 36b
	Measuring	7.00 ⎟		25
C.21	Marking	10.00 ⎟		
C.25	Marking	9.00 ⎟		
	Measuring	5.00 ⎦	191.00X	
	Time-sheet analysis, costings, etc.	236.00		
	Meetings	140.00		
	Supervision and organisation of work	400.00	776.00Y	
	General administration		80.00Z	
	TOTAL		£1047.00	

Total Salary = £5000   Woodland expenditure = 21%

(c) *Office Time-sheet—Analysis*

	£
Time-sheets and wages (forestry)	390.00Y
General office work (forestry)	90.00Z
TOTAL	£480.00

Total Wage = £2080   Woodland expenditure = 23%

Note: X = "direct charge",
Y = "proportional",
Z = "area",
Overheads—see text.

...Arbor...Forest Enterprise
Units £

TABLE 11. *Overhead Costing*

Op. Ref. No.	Compt.	S.H.	Area	Direct Labour Costs	Overhead Direct	Overhead Proportional	Overhead Area Total
			ha	£		(40%)	(×1.78)
1	2	0a & c	10.5	195		78.00	
2	,,	1a	,,	166	0.85	66.40	
3	,,	2d	,,	(18)		—	18.66
4	,,	3a & c	,,	30		12.00	
5	,,	6a	,,	169		67.60	
6	,,	7a	,,	26		10.40	
7	3	7a	5.0	75		30.00	8.90
8	4	23i	20.00	1348	87.00	539.20	
9	,,	23iii	,,	674	44.00	269.60	35.60
10	,,	25	,,	674	44.00	269.60	
11	,,	28	,,	54		21.60	
12						etc.	
13							
14							
15							
16							
17							
18							
19							
20							
21							
22							
23							
24 etc.							
Area with ops.			830.0		428.00	2592.00	
Area without ops.			530.0				
Total			1360.0		428.00	2592.00	

and Total Cost Schedule

Forest year ending 30.9.78

Costs		Total Costs		Unit Area Costs			
Area Per Operation	Total O/H	Direct	Direct and O/H	O/H	Direct	Direct and O/H	Remarks
					£ per ha		
3.11	81.11	514.44	595.55	7.72	48.99	56.72	
,,	70.36	165.70	236.06	6.70	15.78	22.48	
,,	3.11	—	—	etc.	—	—	Contract
,,	15.11						
,,	70.71						
,,	13.51						
8.90	38.90						
8.90	635.10						
,,	322.50						I. 12,320
,,	322.50						E. 5446
,,	30.50						N.I. 6874
1477.00	4497	26436	30,933				
943.00	943		943				
2420.00	5440		31,876				

Key: I = Income. E = Expenditure. N.I. = Net Income.

systems, to allocate the *proportional* overhead costs not proportionally to labour costs but proportionally to time spent on each operation. However, the difference between the two methods is usually not significant and the easiest method has been chosen for convenience in this manual.

The remaining overhead costs refer to the woodlands as a whole and therefore should be allocated on an area basis, the *unit area* category. It must be remembered that the area concerned is the total effective woodland area, unplantable areas such as ponds and rocky outcrops being excluded from the calculation. These costs add up to £2420 in the example and include £80 and £90 (Table 10—Z) as well as employer's liability insurance and all office and miscellaneous expenses of Section B (£169 + £65 and £29), Sections C (£30), D (£34), E (£183) and F (£1740) (Table 9b). The effective woodland area is 1360 hectares and, therefore, overhead costs in this latter section amount to

$$\frac{2420}{1360} = £1.78 \text{ per hectare.}$$

The question arises, when one allocates the overheads on an area basis, as to how the costs should be divided when there is more than one operation per compartment. For example, in Compartment 2, Table 6, there are six operations. Logically, seeing that the overhead cost is on a unit area basis, the cost should be divided equally between the operations, even though the direct costs are different. Therefore, the general rule for an "area" overhead cost is to *divide it equally amongst the various operations within a compartment*.

Woodlands Overhead Costing Schedule

Table 11 gives part of the overhead allocation for the direct operations carried out during the year in question. As can be seen, it includes all three forms of allocation mentioned above, "direct", "proportional" and "per area". This table gives the total overhead charge per operation and the total direct plus overhead charge. Strictly speaking, the only true overhead costs are the fixed costs which have to be allocated on a per area basis; the other "overhead costs" are variable and can be ascribed to the different operations. The compartment operations cards (Table 7) should be posted each year with the appropriate overhead cost, irrespective of whether any operation has taken place in that area. This overhead item is obtained by multiplying the area of the compartment by the unit area overhead cost.

Some of the overhead costs (column 7, Table 11) have been allocated proportionally to labour costs. Actually these overheads include supervision of materials, machines, plants and transport, as well as labour, but it is felt that the latter requires the most supervision. However, if the manager thinks that the whole operation requires equal supervisory time, then he should allocate this group of overheads in proportion to total direct cost and not to labour costs as has been done in Table 11.

Table 11 gives the total overhead per area, per operation for the year in question. This figure could be compared with previous years' results and the trend plotted to give the manager an average figure. Table 11 can also be used to determine the capital valuation of the woods and this will be discussed later.

Headquarter and other overhead costs

It is possible that overhead costs are not confined to a particular project, for it may be part of a much larger enterprise or government organisation. In that case, district, regional, research and headquarter costs will have to be distributed fairly amongst the projects. They should be divided according to the principles outlined above. However, if they are not and the project is confronted with a lump sum, distribution on a per acre basis may be the best solution.

Importance of accurate allocation

Care has been taken in apportioning the costs so that each operation and area reflects a true picture of the overheads. It could be argued that time and effort would be saved if overheads were allocated either on an area or a cost basis, but in both cases misleading results would emerge.

If costs were allocated on an area basis, even though over the whole rotation total overhead costs would be similar, the financial yield would be depressed. This is because over any time period when compound interest is applied, equal yearly sums of money do not give equal returns on investment unless the rate of interest is zero. This will be discussed in greater detail in a later chapter but it can be illustrated with an example from Table 11.

Thinning and felling operations carry the greatest overhead costs because these operations are relatively more labour intensive and require more supervision than other operations, especially as one moves from first thinning to final felling. If overhead costs had been allocated on a "per area" basis they would have amounted to £4 per hectare and not £66 per hectare as is the case for the felling operation (Op. Ref. Nos. 8–11, Table 11). The expenditure for these operations would have been reduced by £62 per ha for this particular year but, assuming other things being equal, it would have had to be spread over previous years' costs. However, the longer expenditure can be delayed, the greater will be the compounded financial return, but the "unit area" overhead method tends to advance rather than delay expenditure on specific areas.

A truer picture may be obtained if overheads were allocated in proportion to costs but again this distorts the results for it emphasises the establishment and harvesting phases and takes no account of areas that, although being managed, have no direct expenditure spent on them. Again, if much work is carried out on contract, allocation on a proportional basis will give a false picture unless account is taken of the contract labour charge as well as the residential labour force. It will also be difficult to judge if contract tenders are lower than "own" costs if overheads have not been allocated properly.

Therefore, although the suggested method is more time consuming, it gives a more accurate picture of and a better insight into overhead spending and could point to the areas where the greatest savings of overhead costs could be attempted.

Interpretation of overhead costs

A certain amount of care is necessary in drawing inferences from the data on overheads. Often they cannot be used to judge the general efficiency of the management. This would apply especially to the results for one year, but they can act as a guide. Many items, such as wet time and transport of labour, depend on the situation and compactness of the land, and so, in the short run, are outside the control of the management. Other overhead items, such as supervision and administration, are of more or less fixed absolute cost; and still others, like expenditure on boundary fences and upkeep and improvement of roads and rides, involve a measure of capital investment. The weight these items give to the overheads total depends in part on the total direct costs during the years in question and they may be influenced by the extent to which an estate in any one year is concentrating on establishment, maintenance or harvesting. A further group of overheads, such as holiday pay, are more or less proportional to wages and thus form a nearly constant percentage of direct labour costs each year.

Again, one big factor influencing the overhead percentage is work done on contract for this involves an increase in direct costs for the estate without the corresponding increase in overheads (Table 11, Op. Ref. No. 3).

It can be seen that overhead allocation in any system of accounting tends to be arbitrary, but with a little practice the manager should be able to apportion the costs with tolerable accuracy. It is important that overheads be allocated correctly because with the large time period involved in forestry a misallocation could make a substantial difference to the profitability of a plantation. The methods set out in this chapter may serve as a guide to the manager. In a similar way, overheads can be allocated to the sawmill and the nursery.

Chapter 7
USEFULNESS OF COSTS

Introduction

Once costs of individual operations and compartments have been assessed, the manager has an important tool at his command which can be used for a multitude of purposes.

Because of the large time involved between planting and harvesting of the crop, the profitability of individual compartments will not be shown in any one year's costing schedule. However, by drawing on costs and returns from other compartments within the enterprise it is possible to estimate the potential yield from each area, by compounding the costs to the end of the rotation, or discounting them back to the start. These methods of determining *financial yield* will be described in Part III. Nevertheless, if compartment records have been kept, straightforward comparisons can be made between compartments to see which area, species or treatment is potentially the most profitable, provided allowance is made for inflation if different years' costs are being compared.

An important task of management is to ensure that all the necessary silvicultural tasks for any given year are carried out in the most efficient manner. This will require the manager to undertake manpower (and machine) planning and it is essential to know the cost of each operation both in money and man-day terms. Cost analysis could help the manager fix piece work and bonus rates and classify the various operations in order of economic as well as silvicultural priority.

The factors of production (inputs) may be varied and the effect on output measured using the marginal analysis technique. Production could be increased by expanding operations or by applying more intensive silvicultural methods. In the former method cost reductions may be brought about by the economies of scale for within certain limits average overhead costs will decrease with increasing operation size. Direct costs, for example fencing, could also be affected in a similar manner (see Chapter 16). Again more intensive silviculture may lead to greater production per unit area. Whether such practices are economic may be determined by marginal analysis.

Stumpage rates—the selling price of the standing timber—could be fixed knowing forest costs and/or forest industry selling prices. Therefore without an adequate knowledge of costs management becomes a difficult task.

Cost Comparisons

Various operations or treatments can be compared and contrasted once the costs are known. For example, different types of planting—tube, notch, turf—could be assessed as well as payment method employed, e.g. day rate vs. piece work rate vs. contract work. Hand draining could be compared to mechanical draining, and the various methods of weeding—hand, chemical, mechanical—critically analysed.

Similar operations could be compared to highlight differences between various compartments, squads of men or size of operation. Again the various operations which constitute ground preparation and establishment could be examined. Ploughing should reduce planting, weeding and beating-up costs; whether in fact this is the case may be determined as well as discovering if this cost is justified.

Costs between different topography, soil types, forests and various parts of a country could provide useful information to a state forest service or a co-operative enterprise for planning and management control. All in all a keen awareness of costs should lead to better planning and increased efficiency.

Piece Work and Bonus Rates

Piece work and bonus payments are ways of increasing the productivity of an enterprise, while at the same time giving a greater reward to the workforce. If the rates are pitched incorrectly they could lead to a dissatisfied workforce and so it is important to determine a satisfactory payment scale. Measured day-rates—the average work done per operation per day—can act as a basis for establishing piece-work rates and these are obtained from the operation costs and time-sheets. Supervision is usually more intensive for piece-work operations but to compensate for this the job is done more quickly and labour additions such as insurance payments may be less. Ideally the rate should be fixed so that both the enterprise and the workforce benefit from the bargain and this will not be discovered unless costs are known. A 25% increase above the measured day-rate may be expected from piece work. Therefore, in theory, the rate could be established somewhere between the two limits. However, the actual rate will have to be fixed by mutual agreement between management and labour. It should be remembered that rates for one specific operation such as planting and thinning may have to be adjusted depending on the type of ground, age and condition of plants and diameter/volume of the trees. Naturally if for thinning the rates are fixed on a number or volume basis, then the rates should vary according to the diameter (or age) of the tree.

In forestry piece work has been successful in easily measured operations such as the length of a fence or drain, the number of plants and the volume thinned or felled. Other operations, for example scrub clearing, weeding and brashing, may be so variable as to make it difficult to fix uniform rates. Again more emphasis may be placed on accuracy rather than speed, making piece work inappropriate. In such cases where increased efficiency is encouraged a bonus system could be introduced, but this would still have to be based on targets and so a knowledge of costs is vital.

Productivity Measurements

Comparisons can be made from year to year and a trend in costs established, thus enabling the productivity increase/decrease to be measured. It is important to compare like with like, otherwise false conclusions may be drawn. For example, if labour productivity for planting is to be measured, the cost per specific number (say 1000) plants is the unit to use. If area costs are compared a false picture may be given for the plant numbers per unit area may vary over time and from place to place.

Similar money units should also be compared: because of inflation and deflation the money unit varies in value over time. Therefore before making comparisons, costs should be in standard units of value or inputs. As a first approximation, or if the rate of inflation is not known, the yearly increase in the minimum wage rate could be used as a deflator.

Productivity may have increased not only as a result of acquired skills or better management but because of changed practices. One operation could directly affect another—ploughing making planting easier—therefore the inter-relationship between operations should be assessed when measuring changes in productivity of a single operation. Differences in establishment costs over time can be compared to give an estimate of productivity changes. However, not only should allowances be made for inflation but actual crop growth should be considered. It may be that such practices as fertiliser application, ploughing and draining, while perhaps adding extra costs to establishment, facilitate an increased crop yield and/or a shorter rotation. All these factors must be considered.

Budget Programming and Control

Once costs are known and labour efficiency measured, the problems of budget programming (and manpower planning) are relatively straightforward. The cost of work for 5-year periods can be made with reasonable accuracy. The time-sheets not only tell the direct costs but give an indication of stoppages such as wet time and illness and therefore estimates can be built into the programme. Cost reduction, a positive management tool, could also be taken into account when compiling the budgets.

Budget programming is a continual activity and therefore the manager may be advised to compile costing schedules covering a shorter time period than a year. A 2-week schedule could be drawn up from the time-sheet shown in Table 2 and this may be used for up-to-the-minute cost comparisons and budget programming.

Previous budgets may be compared to costs, and manpower estimates with task accomplishments. In any project or enterprise one of the most difficult problems is to ensure that the labour force, materials and equipment are being used in the most efficient manner and are always occupied. Detailed costs could help the manager grade operations according to economic (and silvicultural) importance and thus assist in maximising efficiency.

Unit Labour, Material and Machine Inputs—Manpower, Material and Machine Planning

The time-sheets may be used to compile labour schedules for individual and groups of operations, but in place of costs, *unit labour inputs* are calculated, both direct and overhead. The manager can tabulate labour inputs for various operations and for different methods of working, for example hand, mechanical and chemical weeding. Likewise *material and machine inputs* may also be compiled. Once these inputs are known for the various operations and methods of working the manager has a reliable tool for undertaking realistic manpower, material and machine planning. He could also build into his plans possible changes in productivity.

If more sophistication or a time-and-motion study is required, the inputs could be further broken down into actual working, rest time and stoppages.

The three basic inputs may be combined together in different ratios to produce various results. Once the different variations and combinations are known they can be used with effect when undertaking studies such as critical path analysis or budget restraint and when formulating policies both local and national.

The maintenance of a fully occupied stable labour force can be a difficult task but to have the workforce occupied on "profitable" work all the time cannot be achieved without an adequate knowledge of costs and returns.

Marginal Analysis

Standard economic analysis using the theory of outputs, costs and prices (the theory of the firm) demonstrates that profit maximisation will occur when marginal costs—the cost of producing one more unit of output—equals marginal revenue—the revenue from the extra unit of output. (Under "perfect competition" the marginal revenue is the same as the unit price.)

Such analysis could be undertaken in nursery and sawmill enterprises, but it is more difficult in forestry because of the long time gap involved. If all costs and revenues are discounted back to the present day, the point at which marginal costs and revenues are equal will all depend on the discount rate adopted. However, the manager could apply the "financial yield" discount rate to the calculations and work out marginal costs under different production techniques. One obvious technique is to vary the scale of operation or project and see what effect this has on average and marginal cost. Another technique well known in agriculture is to measure the cost of improving yield per unit area. Theoretically there is no reason why yields from forest crops cannot be doubled given the correct strain of trees, suitable ground preparation and fertiliser application. Costs could then be applied to the various inputs and marginal analysis should highlight the most profitable yield.

Stumpage Price

In many forest services the selling price (stumpage or royalty rate) of the standing tree is fixed with very little reference to the growing costs of the crop or the selling price of the manufactured product (for example, sawn timber). In many cases rates have been fixed such that the forest service can never hope to make a surplus on investments. Costs in both the forest and forest industries could be used to arrive at a realistic stumpage price. They may even be used to work out a notional rent or return to the land.

The above are some examples of the use costs may be put to and it demonstrates that although the process may seem laborious, the end results justify the time spent on obtaining accurate costs.

Part II
THE FINANCIAL ACCOUNT

Chapter 8

INCOME AND EXPENDITURE (TRADING) ACCOUNT

Introduction

The method pursued in this manual of placing cost accounting before financial accounting has been quite deliberate. Cost accounting can be said to give a micro-view of the forest whereas a macro-view is displayed in financial accounting. By combining the individual (micro-) costs, both direct and overhead, an *income and expenditure (macro-) account* for the forest estate can be built up. Of course, this income and expenditure (trading) account could be obtained from analysing the cash book, journals and the ledgers but the former method gives the manager a better idea as to where the expenditure went and from where the income came. However, the income and expenditure account only shows the year-to-year running of the estate for no measure has been taken of the increase/decrease in capital value of the forest. Only when this is compiled can a *profit and loss account* be drawn up. Therefore, in order to obtain a profit and loss statement, a *capital valuation* must be undertaken as well. Firstly, however, the income and expenditure account will be described. This account gives a measure of the liquidity of the firm, that is the proportion between the firm's most liquid assets (cash, bank deposits, short-term bills, etc.) and its short-term liabilities (wages, materials, credit purchases, etc.), whereas a capital valuation highlights the strength or weakness of a firm for it is a measure of the total assets and liabilities of the firm.

Methods of accounting may vary but basic principles are described in any standard accountancy textbook. In general the system will include a ledger plus cash, purchase, sales and machine books to record labour payments, internal and external purchases/sales and machine running costs. Separate pages or sections should be opened for unrelated items such as plants and fencing materials and, of course, each item should have its own standard head.

The accounts of the forest enterprise should be kept distinct not only from other sectors associated with an enterprise, such as the farms and shootings, but also from business which is directly connected with the forest, viz. the nursery and the sawmill. Only if this course is followed by the manager will he be able to inform himself and the owner as to the economic state of the individual sections under his care. Table 12 sets out a typical income and expenditure account for a forestry enterprise, illustrated by a worked example. It could form the basis of an accounting system.

TABLE 12. *Suggested Headings and Layout of the Income and Expenditure Account*

TABLE 12a. *EXPENDITURE*

....Enterprise/Year ending ....

The forest enterprise: Income and Expenditure Account   Arbor enterprise
Year ending
30th Sept. 1978

EXPENDITURE	S.H. No.[a]	Worked example £
I. MANAGEMENT SUPERVISION AND LABOUR		10,393
A. DIRECT COSTS (including compulsory Insurance, etc.)		
1. FOREST staff, including working Head Forester and other labour—establishment, tending, harvesting and conversion, and driver or horseman's wages	0–19 22–39 80	6468
2. CONTRACT payments (excluding materials the value of which if not known should be estimated)	S.H. as above	875
		7343
B. OVERHEAD COSTS		
3. Forest staff overhead—repairs and improvements	40–49	16
—protection—fire	50, 51	50
4. Labour additions—wet time, holidays, illness	110–112	532

INCOME AND EXPENDITURE (TRADING) ACCOUNT

TABLE 12a (*cont.*)

	S.H. No.	£		
5. Supervisory work carried out by the Head Forester	20, 21 & 120	325		
6. Salary or proportion of Woodland Manager	121	1047		
7. Salary or proportion of office staff	124	480		
8. National Insurance of 6 and 7 if not included above (incl. graduated pension contribution)	125	Incl.		
9. Owner: value of services (if any)	122	Nil		
10. Director's and/or Consultant's fees	123	Nil		
11. Pensions and contributions to private pension schemes	114, 116, 126	139		
12. Employer's liability insurance	115, 126	9		
13. Perquisites in kind (all forest enterprise personnel)	117, 129	308		
14. Vermin control (a) by gamekeeper,	55	Nil		
(b) by vermin eradication soc./organisation	56	24		
			2930	
C. *FOREST LABOUR HIRED TO OTHER DEPARTMENTS*				
15. Sawmill, nursery, etc.	60's & 70's	120		
			120	
				12,073
II. *VEHICLES, MACHINES, TRANSPORT AND TOOLS*				
1. Vehicles, machines and plant (excluding buildings)			1646	
(a) Depreciation	85	289		
(b) Repairs	83	120		
(c) Fuel, oil, lubricants, animal feed	81, 82	1197		
(d) Insurance, licence, etc.	84	20		
(e) Profit (add)/Loss (subtract) on machine account		20		
2. Hire of vehicles and machines for internal use:			10,397	
(a) From other estate enterprises	86	41		
(b) From outside (contractor's charges, etc.)		10,356		
3. Carriage in or out (external transport costs)	87		Nil	
4. Tools			30	
(a) New	88	26		
(b) Repairs	89	4		
III. *CAPITAL EXPENDITURE DEPRECIATION*			Nil	
1. New roads and rides— depreciation	44/46	Nil		
2. Others—depreciation		Nil		

TABLE 12a (cont.)

	S.H. No.	£	£
IV. CONSUMABLE STORES, ETC.			7077
1. Fencing material from sawmill and from outside	90, 91	2263	
2. Draining material except for road construction	92	Nil	
3. Roads and rides material for maintenance	93	Nil	
4. Plants from nursery and outside	94, 95	4285	
5. Fertilisers for plantations	96	520	
6. Protection—weed killer, pest killer, etc.	97, 98	4	
7. Fire fighting equipment	99	5	
V. BUILDINGS AND LAND. RENTAL VALUES AND RATES			2233
1. Rent and rates			493
(a) Forest workers	118	420	
(b) Manager (all or part)	127	63	
(c) Office (all or part)	130	10	
2. Land and Taxes			1740
(a) Rent	140	1700	
(b) Land tax	141	40	
(c) Product tax	142	Nil	
VI. ADMINISTRATIVE EXPENSES			220
1. Travelling			36
(a) Transport of workers to and from work	113	Nil	
(b) Cycle allowance	113	Nil	
(c) Travelling expenses for Manager/Owner	128	36	
2. Office expenses	131/133		55
3. Audit and legal fees	134		9
4. Subscriptions to societies (excel. vermin clearance)	135		10
5. Insurance premiums			105
(a) on office, house, public liability, etc.	136	5	
(b) Woodlands	54	100	
6. Bank charges, including interest on overdraft	137, 143		2
7. First aid, bad debts, miscellaneous	138		3
TOTAL EXPENDITURE			£31,996

[a]S.H. No. = Standard heads, these numbers are only included to act as a guide.

Expenditure

Labour

The expenditure side has been divided logically into a number of sections. The first—*management supervision and labour*—contains all the labour costs of the forest enterprise during the year. For convenience it is split up into three subsections—*direct costs, overhead costs* and *hire of labour*. The total for each group can be arrived at in two ways, either from the direct costing and overhead schedules (Tables 6 and 11) or from the wages and salary sheets. The direct costs are simply the totals of the enterprise and contract labour for the year in question. Likewise, the overhead costs can be extracted from Table 9 (or 11), being careful only to include the labour costs or their equivalent (e.g. perquisites which are treated as payments in kind). A check with the standard head numbers given opposite each entry should sort out the difficulties. For example, entry No. 11 (Table 12)—pensions and contributions to private pension schemes—includes an item from the labour additions section of £104 (Table 9b) plus £35 from the next section—Administrative expenses.

Finally, in the management, supervision and labour section there is a subsection for the forest labour hired out to other departments or to outside. This is usually a straight cross-entry, the income received from hiring out the labour being exactly the same as the labour charge. For example, in Table 12 the forestry squad worked in the sawmill during wet weather and cut firewood. The expenditure and income, debited and credited to the forest enterprise, is exactly the same (£120) and this sum, if added to the forestry costs (Table 11), will equal the expenditure total (Table 12). This sum is just a small item out of the total wage bill of the forest enterprise. Therefore, no overheads have been allocated to this figure. (In fact, to be strictly correct overheads of about £48 should be added to this wage bill.) If labour hire bulks large in the accounts then this item should include both direct and overhead labour costs. Again, if the labour is being hired out for non-enterprise purposes the manager will quite rightly include, as well as charges for direct and overhead costs, a profit margin. In a case like this the expenditure and income sides will not show the same figure.

Machine costs and depreciation

Section II of the expenditure side (Table 12) covers the use of machines, purchase of tools, hire of vehicles and carriage in or out. The first subsection—the running costs of the enterprise machines—can be obtained from the log book, a summary of which is given in Table 13.

Table 13a, the Machine Account Summary, lists the various machines

TABLE 13. *Machine Accounts*
TABLE 13a. *Summary of Machine and Running Cost Account 1978*

		Opening valuation	Closing valuation	Depreciation
A.	Power saws—Depreciation rate 25% p.a. of cost			
	Year of purchase	£	£	£
	P. 77	70	52	18
	P. 76	42	32	10
	P. 75	42	32	10
	P. 74	24	18	6
	P. 73 (scrap)	20	—	20
		198	134	64
	Repairs and maintenance and spares			75
	Petrol and oil			880
	Total expenditure			£1019
	Total expenditure as charged to the forestry enterprise for use of saws			£1036
	Difference		+	£17
B.	Tractor and winch—Depreciation rate 28⅛% per annum of cost			
	P. 76	800	575	225
	Repairs and maintenance and spares			45
	Petrol and oil			317
	Licence; Insurance			20
	Total expenditure			£607
	Total expenditure as charged to the forestry enterprise for use of tractor and winch			£610
	Difference		+	£3
	Correction to be credited to the vehicles and machines—expenditure section		+	£20

INCOME AND EXPENDITURE (TRADING) ACCOUNT

with their value at the start and end of each financial year. A standard depreciation rate is applied to each group of machines. The particular rates, which can be obtained from the government's taxation officers, are shown in the table, together with the depreciation allowance.[1] The money spent on repairs, maintenance, spares, petrol, oil, licence and insurance is also listed and the total expenditure is given for each group of machines.

Table 13b details the capital account for the machines. The depreciation of all the machines appears on the expenditure side of the income and expenditure account (Table 12 (II, 1a)), but it must be included as well in the above account (Table 13b) as a balancing item.

TABLE 13b. *Machine Capital Account 1978*

	Start of year	End of year
Machine valuation (including purchases during the year)	£998	£709
Depreciation		£289
	£998	£998

An enterprise may have other items of capital expenditure, such as road construction, house building or the erection of boundary fences. In such circumstances separate accounts should be opened similar to Tables 13a and 13b.

If a comparison is to be made, say between the power saws in order to compare one make or machine with another, then the log book should separate out all the costs relating to the different machines and a check made of hours run. In this way individual machines can be compared with the same, or different models.

It will be noticed in Table 13a that there is a line for the total expenditure as charged to the forestry enterprise. When machines have been costed at a fixed rate per hour it is certain that the amount charged for the machine will not be exactly the same as the actual running cost.

---

[1] As mentioned in Chapter 3, standard rates have to be applied for external accounts, but for internal purposes these rates should be realistically tied to the working conditions. Inflation has not been included in the financial accounts because it may or may not be allowed as a legitimate cost by the taxation authorities.

Therefore, if the total expenditure of the machines as charged to the forestry enterprise is more than the running costs, then the difference is added to both the income and expenditure sides of the account. On the other hand, if the total expenditure is less than the running cost, the difference should be subtracted from both sides of the account. If the costing of the machines is to be entirely accurate, then the charge as listed in the costing table (Tables 6 and 8) can only be made after the financial year has closed, when all the cost elements are known and allocated. However, if the costs are based on the previous year's results, the manager can obtain costing details immediately with very little loss of accuracy. This method has been used in the costings section (Table 6) and therefore an allowance has to be made in the income and expenditure account to correct this slight inaccuracy. For 1978 the correction amounted to plus £20 (Table 13a).

The subsection in this group (Table 12 (II, 2)) covers the hire of vehicles and machines from the enterprise or from outside. Naturally this includes the contractor's costs (£10,356) (excluding labour), and these are shown together with the cost (£41) of transporting the fencing materials and the plants, which it is assumed have been transported by an estate machine not belonging to the forest enterprise (see also Table 6). Should the forest enterprise own a vehicle then the running costs must be entered in the machine account (Table 13) and the transport costs would therefore appear in the previous subsection.

Carriage in or out, Table 12a (II, 3) is for the transport cost of materials to be used directly on the forest enterprise; for example, the charge of plants transported by train, or the cost of a lorry to take Christmas-trees to the market.

Tools, Table 12a (II, 4), are generally a small item that are not worth depreciating, hence the whole cost is credited to the year of purchase. The distinction between a tool and a machine can be arbitrary at times, but usually tools do not have any running costs, just repairs.

Capital expenditure—roads

The third section on the expenditure side is for capital depreciation of new roads and other items. Capital depreciation has been discussed already in the chapter on "Overheads" but perhaps it would be advisable to give an example since there is no expenditure credited to this section in Table 12a. Let us assume that new road-work has been carried out on an enterprise for the past 5 years and previous to this no new road-work had been undertaken for more than forty years. Table 14 is a summary from the Road Account of this particular enterprise.

TABLE 14. *Summary from Roads Account 1978*

Year	Kilometres completed	Cost	Miscellaneous
1974	1	£ 4000	Contract
1975	3	10,000	,,
1976	3	8000	,,
1977	4	12,000	,,
1978	—	—	—

Year	Value at start of year	Depreciation (rate 1/40th)	Value at end of year
1974	£4000	£100	£3900
1975	13,900	348	12,652
1976	21,652	541	21,111
1977	33,111	828	32,283
1978	32,382	807	31,476

In 1978 in this particular enterprise £807 will be entered in the capital expenditure section, Table 12a (III), as depreciation on the new roads. If, at the same time, a sum of money had been spent maintaining the roads this should be credited to the next section—consumable stores, Table 12a (IV). In theory the construction of woodland boundary walls, fences and dykes are items of capital expenditure and should be depreciated. Also, the results of ploughing and draining should last more than one rotation and therefore these items of expenditure are strictly of a capital nature. One can only plough an area of land, except with difficulty, before it has been planted for the first time, and it is essential that initial draining be carried out properly so as to prevent windthrow and costly redraining. However, for the sake of convenience, these last two items are credited to the direct cost of establishment of plantations, but the excavation of main drains and erection of boundary fences, etc., can, if the manager has detailed costs both past and present, be put in the capital account and depreciated, otherwise they will appear as overhead items.

Consumable stores (Section IV)

This section includes purchase of fencing materials (including items provided by the forest enterprise itself), plants, fertilisers, etc. All these items should be recorded in the accounts book and can be extracted from it.

Rental, rates and administrative expenses (Sections V and VI)

The final two sections of the expenditure side—buildings and land, rental value and rates (V), and administrative expenses (VI)—are self-explanatory and need no further elaboration. In the worked example these items have been extracted from the overhead schedule (Table 9).

Income

Timber sales

The income side of the accounts is again divided into a number of convenient sections, the first being for the sale of timber. It includes sales to outside buyers, timber for internal use such as fencing posts, and sales to the organisation outside the forest (the enterprise sawmill). A separate letter A, B, C, can be used to denote sales to the different buyers mentioned above. As pointed out previously, a realistic value must be placed on timber sold to the enterprise sawmill so as not to distort the economic picture of either account.

Minor forest products

The second section is for the sale of minor forest products and includes income from scrub-clearing activities as well as the estimated value of free supply of firewood (if any) from the forest to the members of the enterprise.

Government grants, tax rebates and miscellaneous receipts

Grants and tax rebates are entered in the third section. In certain countries, for example the United Kingdom, tax rebates can be claimed

INCOME AND EXPENDITURE (TRADING) ACCOUNT

TABLE 12 (*cont.*)
TABLE 12b. *INCOME*
The forest enterprise: Income and Expenditure Account

Arbor enterprise
Year ending
30th Sept. 1978

INCOME

		S.H. A.B.C.	Worked example £
I.	**SALES OF TIMBER**		50,052
	1. Unconverted:		46,978
	(a) standing	200	
	(b) blown	201	
	(c) at stump	202	
	(d) rideside	203	
	(e) delivered	204	46,978
	2. Converted:		3074
	(a) poles	205	
	(b) fencing material	206	
	(c) decorative material	207	
	(d) commercial wood	208	3074
II.	**OTHER SALES OF FOREST PRODUCTS**		39
	(a) Bark	210	
	(b) Firewood (sold during scrub clearance)	211	12
	(c) Cut firewood in woods (including free supplies to estate employees including forestry workers)	212	
	(d) Christmas-trees	213	24
	(e) Miscellaneous	214–219	3
III.	**GOVERNMENT GRANTS AND TAX REBATES**		Nil
	(a) Grants and loans	260	
	(b) Tax rebates	261	
IV.	**MISCELLANEOUS RECEIPTS**		140
	(a) Rents	264	
	(b) Hiring out of labour to other depts.	266	120
	(c) Hiring out of vehicles and/or machines	267	
	(d) Other receipts	265 & 268	
	(e) Machine log book—Profit (add)/Loss (subtract) on machine account		+20
	TOTAL INCOME		50,231

SURPLUS/DEFICIT

Income	£50,231
Expenditure	£31,996
Income minus expenditure	£18,235

against losses on certain parts of the woodlands, hence the reason for its inclusion in this section.

The final section of the income side is for miscellaneous receipts and includes hiring out of labour, vehicles and machines, as well as the machine log-book correction (Table 13).

Surplus/Deficit for the year

A summary of the income and expenditure for the year in question is given in the final section and the surplus or deficit for the enterprise may be calculated. Only if the forest has a uniform age class distribution will the year to year surplus (deficit) give a true reflection of the profit (loss) of the enterprise. Then it can be anticipated that a constant volume of timber, by size class, will be forthcoming from the woods. This concept of normality is rarely achieved, indeed it may not be the aim of the manager for he should be flexible enough to adjust the management according to the demands of the market. However, the manager may obtain an approximate idea of the profitability of the enterprise by comparing *actual yield* to anticipated *average yield* or looking at the *present and "normal"*[2] *age class distribution*.

Degree of profitability

Natural woodlands may give an average yield as low as 1 to 2 m^3/ha/year of commercial timber whereas in plantations the anticipated yield may be between 10 and 20 m^3/ha/year. Again for plantations, approximately 35–40% by area should be in the young immature stage, 55–60% in the productive phase and the remaining 5% in the reserve of plantable land. If the percentage of woodland in the formation stage is greater than

---

[2] The term a normal forest is often used in forestry as a standard by which to compare other forests. *The British Commonwealth Forest Terminology* (Anon., 1953) defines a normal forest as follows:

"A standard with which to compare an actual forest so as to bring out its deficiencies for sustained yield management; a forest which for a given site and given objects of management is ideally constituted as regards growing stock, age class distribution and increment and from which annual or periodic removals of produce equal to the increment can be continued indefinitely without endangering future yields."

Therefore, in the above context a normal age class distribution is an equal representation of every age class.

INCOME AND EXPENDITURE (TRADING) ACCOUNT 91

average, the surplus should be less than normal. On the other hand, if the volume removed is larger than normal then the surplus will be exaggerated. By examining the age class distribution and/or yield, the degree of profitability can be explained and estimated. However, in order to draw up a true *profit and loss account* a capital valuation of the woodlands must be undertaken.

Chapter 9

CAPITAL VALUATION

Introduction

Capital valuation as the name implies is the valuation of a firm's capital. However, in forestry capital valuation has a special meaning because the tree is regarded as both the (standing) capital and the (felled) product. Therefore, the term capital valuation usually refers to the valuation of the growing stock. In normal accounting practice this would be regarded as stock valuation but trees are rather distinct because of their power of growth and the long time interval involved between planting and felling. Thus whereas in manufacturing firms, there is a quick turn over of the product and the stock is only a fraction of the yearly production, in forestry the opposite is the case. At any one time there is a high ratio of growing stock (inventory) to annual production. It can vary from about 5 to 1 for some fuelwood/pulp plantations to 50 to 1 or more in some fine hardwood forests. Account should therefore be taken of this appreciating capital as well as the inanimate capital such as road, buildings and machines; this latter has already been dealt with under the income and expenditure account-depreciation of vehicles, machines, roads etc.—and this chapter deals with the valuation of the growing stock.

Capital valuation of the growing stock is important for a number of reasons; besides determining the liquidity of the firm which can be assessed from the income and expenditure account it is expedient to discover the soundness or solidity of a firm. The soundness of a firm is measured by the relationship between the total assets and liabilities and this includes a valuation of the firm's fixed assets such as buildings, land and forests. Capital valuation gives a measure of the worth of a forest at any specific moment in time and this may be useful when buying or selling a forest property or dividing it amongst heirs, when evaluating the forest property as a security against loans, when assessing the property for the purpose of insurance and taxation, when assessing the value of part of the forest to determine compensation for fire damage or expropriation and, finally, when entering the value of the property on the balance-sheet.

If capital valuations are undertaken annually it is easy to determine the amount by which the crop has increased or decreased in value during the year. This amount can then be added to (or subtracted from) the surplus/deficit as given in the income and expenditure account and a profit and loss account may be compiled for the year. Naturally if a forest is in an immature phase one would expect the capital to be increasing each year thus making a positive contribution to the profit and loss account. On the other hand, if the forest is mature the capital could be decreasing as "excess" thinning and felling takes place. This decrease should be taken into account in the statements. Only when there is no increase or decrease in the capital value of the growing stock will the income and expenditure account be the same as the profit and loss account.

Subjective Nature of Valuation

The valuation of a forest area is based on a number of objective measurements such as the area, age, volume and productivity of stands. However, the valuation may involve a number of assumptions which are more or less subjective, for example rotation age, future costs and prices. Again valuation depends on how the forest is going to be used— managed on a sustained yield basis or exploited as quickly as possible and then sell the land or change its use. The value of a forest may depend on who is the potential buyer. If the potential buyer already owns a forest property, additional forest land may be of high value for it may enable the owner to make fuller use of existing machinery, administrative capacity, etc. A similar argument could be applied to the owner of a forest industry for such a forest would guarantee the supply of raw material.

Forest valuation is neither exact nor objective; valuation is a subjective estimation because any value depends on the person who is assessing and the intentions as to the use of the object. However, certain methods are accepted in accountancy and these methods will be discussed.

Methods of Capital Valuation

Capital valuation is necessary in order to show the profit or loss earned in a particular year. This profit (or loss) may affect the tax payments due (or refunded), therefore, the valuation method has to be acceptable to the taxation authority as well as being fair to the individual. The first method, and the one favoured in this manual, endeavours to assess the *actual value of the forest in its present state*; it does not consider the potential value but only its actual value by means of a stock valuation.

There are two main ways of stock valuation, one using costs and the other using prices. The first way is considered the usual or "normal" method whereas the second is only used in special cases such as tea, rubber, forest and mining enterprises. Normally stocks are valued "at cost" *or* the lower of "at cost" or "net realisable value" (value from disposal of stock less all expenditure) or "replacement price". Such a system is suitable for manufacturing industries where the stocks are inanimate and have not the power of growth. If we consider stocks that are growing, such as agricultural/forestry crops, the "at-cost" valuation system is not appropriate for it does not take into account the dynamic nature of the stock; this is why the "selling price" method is more suitable. However, in forest crops there is a distinct time interval when the crop is immature and has no value as such, only a potential value. In this phase it is probably more appropriate to value the crop "at cost", but also to make some allowance for its growth. This is the method recommended, that is value the established crop at cost, increase the value yearly to account for increment in the non-marketable phase and, as soon as the crop becomes merchantable, value it at its market price. An example is worked out demonstrating this method of valuation. However, this "actual value" method may not be favoured by the forester or be acceptable to the taxation authorities in individual countries, therefore other methods are described. These include the *realisation value*, which is like the first method but only takes into account saleable timber. A third method, *the potential value*, examines the expected yield (the financial yield) from a plantation and applies this yield rate to the net invested capital. A variation of this method is called the *expectation value* and this is discussed. *The capitalisation value* method assumes that the forest will be managed in perpetuity and will reach a state where there will be an equal distribution of all age classes and an equal annual volume production. The forest investment should then produce an equal annual income, which can be capitalised. Finally, the *"Accountants" capital valuation* will be discussed. This only considers actual investment.

Forest Valuation using the Actual Value Method

Forest valuation may be divided into three parts, the valuing of the mature area, that is plantations containing merchantable timber; secondly, the assessing of the immature areas; and, lastly, the valuation of the bare land. A distinction is made in this text; firstly, between land capable of growing a crop and unproductive areas—rock outcrops, waterlogged areas, ponds and land above the tree line. Secondly, plantable land is divided for convenience into three, namely, bare land, planted or naturally regenerated land which has not yet reached the merchantable stage (usually the first commercial thinning but sometimes earlier if trees are raised for greenery or Christmas-trees) and, lastly, trees that have reached the merchantable stage which includes all phases from the first commercial thinning to the final felling. These three stages will be referred to as *bare land, immature crop/area* and *mature or merchantable crop/area*. Figure 2 illustrates the three valuation stages.

Before planting, the only value is that of the land. Between planting and the first commercial thinning the crop has only a potential value but capital investments have been made as indicated by the actual investment line. The valuation during the immature stage is assumed to be the actual investment plus a percentage to account for the growth of the crop. The actual percentage rate to add should be such that the value just before the first thinning is marginally less than the value of the crop after first thinning. Finally, once the crop becomes merchantable it can be valued at the going market price because theoretically all the crop could be liquidated at any moment of time. It will be seen that the crop valuation does not increase uniformly from year to year; it

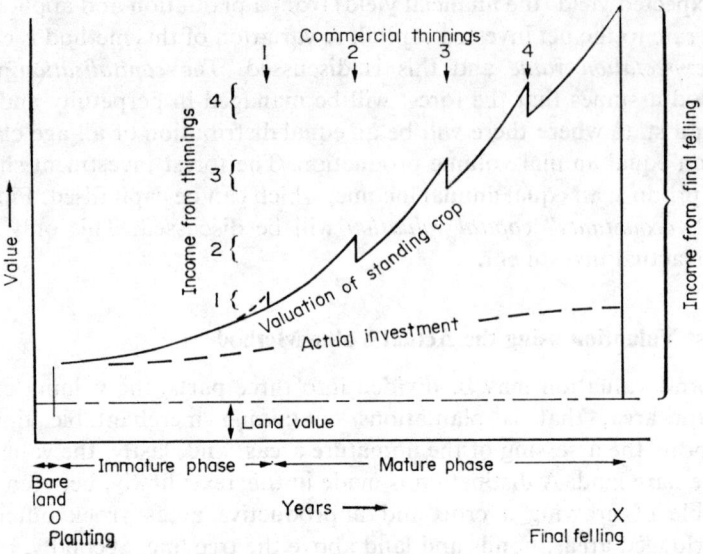

Fig. 2. Capital valuation forest plantation.

starts off slowly then rises rapidly and has a number of peaks and depressions caused by the removal of some of the crop as thinnings. After final felling the value of the area reverts to the value of the land.

Figure 2 can be looked on either as a single crop proceeding through the various stages of growth or as the value of a number of compartments at different stages of growth, each compartment having a similar area and soil type. The following valuation looks at a single crop proceeding through the various stages of growth, but it can be taken as a number of compartments at different ages. The valuation starts with the merchantable phase for it is easiest to explain and this valuation is required before the immature phase valuation can be determined.

Unit Area Valuation of the Merchantable Crop

To value the merchantable area of a forest it is necessary to discover the standing volume or age and yield class of each compartment. Once this is known, a valuation can be placed on timber using current prices. Each subsequent year the value can be increased to account for volume, increment, and unit price changes.[1] If yield tables are used the unit area volume of the main crop after thinning is usually given together with the yield from thinning. It is therefore easy to value the main crop at thinning time by taking the volume figures as given. In order to value the crop between thinnings just add each year, one-third (or whatever the thinning interval is) of the difference between the main crop volumes. An allowance should also be made for the increment in thinning volume; this starts from zero at the time of thinning and rises to a peak just before the next thinning (see Figure 2). Again knowing the thinning interval and the thinning volume, an equal volume, say one-third of the total (for a 3-year thinning cycle), may be added each year and then priced accordingly.[2, 3] It is important not to value the thinning volume in the year of thinning for this value will be recorded in the income and expenditure account as a sale, and therefore the capital value has been decreased by this amount. The method of valuation is illustrated in Table 15b(ii), *capital value of the merchantable area*, and value from the thinnings and main crop up to year 55 is shown in Table 27, Chapter 12. Likewise when an area is clear-felled its book value is subtracted from the capital valuation of the forest (Table 16b). Similarly if a selection felling takes place the capital valuation is reduced in proportion to the area felled. It is important to keep under review the book value of the woods and to revise them up or down when a price change takes place. This is not to say that every price change must be accounted for because many may be seasonal, but yearly changes should be recorded, consequently the book value of the crop will be the same or very similar to the actual net value when the crop is felled. A record should be kept of prices (by size class) over the years and the trend established. If home timber prices keep at par

---

[1] It is possible for the value of a stand to decrease temporarily if unit price decreases more than volume increment.

[2] Strictly speaking the volume does not increase by an equal annual amount. More accurately it increases by an equal percentage but the former assumption will not lead to any great error.

[3] To save time with little sacrifice of accuracy the value of the thinning volume between thinnings could be neglected.

with other commodity prices then the trend should keep pace with inflation.[4]

Valuation of Young Woodlands

When the valuation of immature woodlands is considered a number of problems spring to mind. Should the plantation be valued at average establishment or replacement cost, and what figure should be used—the net or gross sum? What allowance should be made for increase in value of the plantation as it grows older and approaches the merchantable stage? How does one value a plantation? Does one use current costs or actual historic costs? There are a number of answers to the above problems, and the woodland manager may disagree with the solution chosen in this manual. Therefore, other methods of valuation are also discussed.

Replacement cost can differ from establishment cost because certain establishment operations need not be repeated, for example ploughing and draining. If a fire destroyed a plantation the only establishment costs involved may be planting, weeding and beating up. Therefore, one would normally expect the replacement value to be less than the establishment cost and the sum to vary depending on the work involved. However, if there has been a fire and destroyed the crop an element of time has been lost which cannot be replaced and this could be treated as a cost. Again, does one use gross or net costs? If, in ground preparation, income is obtained from selling scrub as firewood this income is subtracted from the gross preparation costs to give a net figure. Should one do likewise with government grants and tax rebates? Some enterprises may not qualify for government grants and/or rebates and some countries may not give them. However, to be consistent the logical valuation figure to choose is the *net establishment* cost, for this is the actual net cost to the enterprise. Again if governments are undertaking afforestation wholly or partially on social grounds, some or all of the labour costs should be discounted.

[4]On the basis of historical records of timber prices in the U.K. and in the international market, it has been estimated that the real price of timber might be expected not only to keep pace with other commodities but rise at a rate of $1\frac{1}{4}\%$ per annum, at least in the U.K. See *Forest Planning* by Bradley, Grayson and Johnston, p. 151, appendix II, Faber & Faber.

Unit Area Valuation of Immature Woodlands using Actual Costs and Prices

If detailed costing records have been kept for the compartments (Table 7)[5] these figures may be used for valuation, provided allowances are made for general inflation,[6] and the increase in value of the growing stock. What should the allowances be? The general inflation rate varies from country to country and from year to year but average figures are published by governments and banks for individual countries and these are reliable rates to use. As for the increase in the value of the growing stock, an average uniform rate of increase should be chosen so that the assumed value of the stand just *before thinning is nearly equal* to the *assessed value* of the main crop after first thinning. The valuation of the forest at first thinning depends on a number of things—the yield class, the species, average price, percentage stocking, etc.; it may be so low that the establishment and tending costs are in excess of the valuation. In a case like this the forest should be considered to be non-productive until both valuations are similar. However, it is possible that the establishment and tending costs are always in excess of the "selling price" valuation of the forest or the selling-price valuation is just greater than the incurred costs only at the end of the rotation. If this is so then it is extremely doubtful whether the investment should have taken place at all. As an economic proposition it is a non-starter but there could be other over-riding goals such as soil and water conservation, amenity and recreation, etc. Then values should be imputed to these and other multi-purpose activities so that a clear idea is obtained of the cost or value of each activity.

It is assumed that before investing in a particular forest project or enterprise, a project appraisal has been carried out and the decision to proceed taken knowing that the anticipated return is acceptable to the investor and is at least as good as in the next best alternative. In such

---

[5] The compartment record cards should include details of government grants and tax rebates, in fact all income as well as direct and overhead costs.

[6] Two kinds of inflation are recognised in this book: (a) general (money) inflation —the decrease in the value of money; and (b) relative inflation—the increase in the value in the real price of timber when compared to the average price of other commodities. For further information on this latter point see *Forest Planning* by Bradley, Grayson and Johnston, p. 151, Faber & Faber.

cases the net capital cost—that is capital cost less any receipts—should usually be less than the forest valuation when the timber reaches merchantable dimensions. Table 15a shows how to value a particular compartment from establishment to just before it enters the merchantable phase. Here historic costs have been used. First they have been

TABLE 15a. *Capital*

(a) *Using historic costs* Land value 1961 £62 per ha.
(i) *Immature area*
Summary of unit costs taken from the record card (Table 7)
Compt. 7. 40 ha. P. 1961. 1st Thinning 1978
Units of costs and receipts are £/ha

Year	Age	Operation S.H. nos.	Costs Direct	OH	Total net cost[a]	Cumulative cost	Inflation multiplier for 1978[b]	Historic cost at 1978 value[c]
1961	0	0 to 7	85.00	15.00	100.00	100.00	1.615	161.50
62	1			0.71	0.71	100.71	1.583	1.12
63	2	7 & 8	3.00	1.73	4.73	105.44	1.552	7.34
64	3			0.75	0.75	106.19	1.522	1.14
65	4			0.77	0.77	106.96	1.492	1.15
66	5			0.79	0.79	107.75	1.463	1.16
67	6			0.81	0.81	108.56	1.434	1.16
68	7			0.83	0.83	109.39	1.406	1.17
69	8			0.85	0.85	110.24	1.379	1.17
70	9	10	2.50	1.17	3.67	113.91	1.351	4.96
71	10			0.89	0.89	114.80	1.325	1.18
72	11			0.92	0.92	115.72	1.299	1.20
73	12			0.94	0.94	116.66	1.274	1.20
74	13			0.98	0.98	117.64	1.249	1.22
75	14			1.02	1.02	118.66	1.224	1.25
76	15			1.10	1.10	119.76	1.177	1.29
77	16	17	22.73	5.00	27.73	147.49	1.100	30.50

(ii) *Merchantable area* (see also Tables 25, 26, 27)

Year	Age	Operation S.H. no.	Main crop after thinning		Value per m³ (£)	Capital valuation (£/ha)
			Stems/ha (No.)	Volume/ha (m³)		
78	17	22 (1st)	1750	100	3	300

CAPITAL VALUATION

tabulated on a unit area basis, the direct costs being kept separate from the overhead costs for convenience. The receipts if any are subtracted from costs so that total net costs per year and cumulative costs may be calculated. It will be seen that the cumulative costs increase from £100 in year 0 to £147 in year 16 and this is less than the capital valuation of

Valuation per Unit Area

Cumulative "1978" cost	Increase in value to account for growth of the capital				
	Value at year's start	Increase in value during year (2.2% per year)[d]	Year's cost	Value at year's end	Capital valuation
161.50	0	0	161.50	161.50	162
162.62	161.50	165.05	1.12	166.17	166
169.96	166.17	169.83	7.34	177.17	177
171.10	177.17	181.07	1.14	182.21	182
172.25	182.21	186.22	1.15	187.37	187
173.41	187.37	191.49	1.16	192.65	193
174.57	192.65	196.89	1.16	198.05	198
175.74	198.05	202.40	1.17	203.57	204
176.91	203.57	208.05	1.17	209.22	209
181.87	209.22	213.82	4.96	218.78	219
183.05	218.78	223.60	1.18	224.78	225
184.25	224.78	229.72	1.20	230.92	231
185.45	230.92	236.00	1.20	237.20	237
186.67	237.20	242.42	1.22	243.64	244
187.92	243.64	249.00	1.25	250.25	250
189.21	250.25	255.76	1.29	257.05	257
222.76	257.05	262.70	30.50	293.20	293

Notes

[a]Total net cost = Gross cost less receipts (if any). Such receipts include Sale of scrub for firewood, Government grants, Tax rebates, etc.

[b]The following inflation rates have been assumed:

    2 per cent per annum   1961–1974
    4 per cent per annum   1975
    7 per cent per annum   1976
    10 per cent per annum   1977

[c]Values given at the start of 1978 or end 1977, etc.

[d]Increase in value to account for real growth of capital 2.2% per year, appropriate rate. Range: Upper limit 2.3%. Lower limit 1.8% (see Example 10).

the main crop (after thinning) estimated to be £300 at age 17 (Table 15a). That is to say, if the crop was clear felled at the age of 17 years it would be worth £300 per ha plus whatever money was realised from thinnings (£50). The crop *after* and not *at* thinning time is valued because the object of the exercise is to assess the value of the capital. Thinnings (and final felling) are not capital but disposed stock and their value is recorded as sales in the income and expenditure account. Of course it is advantageous not to fell at 17 years but allow the main crop to grow to maturity, for its potential value is greater than its actual value (see Table 27), but if for some reason the crop had to be felled or assessed for insurance or fire damage then this valuation would give a true picture of its present worth.

The crop when it has reached the merchantable phase (age 17 years) is valued at £300 and the cumulative capital expenditure just before the productive phase is £147 (age 16 years). However, this capital expenditure is recorded not in constant money values but in changing values for no account has been taken of inflation.

In order to account for inflation the average inflation rates have to be added to each year's costs. Thus for 1973, 10% has been added to the cost. Likewise for 1972 the compounded value of 7% multiplied by 10% (17.7%) has been added to the year's cost and so on. The average inflation rates assumed for the particular years are stated in Table 15a and the inflation multipliers are given in column 8. Thus column 9 gives the historic (or actual) costs in 1st January 1978 (constant) value terms and column 10 gives the cumulative 1978 constant value costs. The 1961 initial establishment and weeding costs have been increased from £100 to £162 to account for inflation; the total cumulative costs are £223, an increase of £75 over actual costs. The "inflated" cumulative cost at year 16 is still less than the capital valuation in year 17. The difference is because the capital is not static but growing. Therefore, to account for the dynamic nature of the capital a percentage has to be added each year to the capital so that the value of the capital at the end of the immature phase is just less than the capital valuation at the start of the productive phase. In the example (Table 15a) the appropriate rate to account for the real growth in capital is 2.2% per year. However, this rate will vary from site to site and more dramatically from country to country. It is to be expected that tropical countries with relative low labour costs and shorter rotations would have cumulative capital costs at the end of the immature phase considerably less than the capital valuation at the start of the productive phase. Here the appropriate rate to account for the real growth in capital may be as high as 10%; it all depends on costs, prices, yield and rotation.

The appropriate rate for each project site class or species can be arrived at by trial and error with the help of compound interest tables and/or desk calculators once the cumulative costs and main crop value are known.[7] However, the limits within which the rate falls may be calculated by knowing the initial capital investment (I), the cumulative costs (C) and the standing value of the crop in the year of first commercial thinnings (S). The difference (D) between the standing value (after thinning) and the cumulative costs is the value that has to be spread proportionately over the immature phase costs. The upper limit is fixed by taking the above difference (D) and working out its percentage of the initial capital investment (I). This "upper" percentage is then discounted by the number of years to first commercial thinning (n) to find out the annual percentage increase. Likewise the lower limit is fixed by taking the difference (D) and determining its percentage of the cumulative costs (C) and then working out what this percentage increase is on an annual basis. These two annual percentage figures form the limits in which the appropriate percentage must lie. Whether it is nearer the upper or lower limit depends on the spread of capital investment over the immature phase period. If most of the investment, after initial establishment, took place early on, then the appropriate rate will veer towards the lower limit. On the other hand, if most of the investment took place just before first thinning, as is the case in Table 15a, then the rate will be near the upper limit. An example, Example 10, using the figures given in Table 15a, will illustrate how the range bounding the appropriate rate is fixed.

---

[7]Another method, simpler without much loss of accuracy, would be to take the difference between the cumulative cost (after inflation) just before first thinning and the main crop capital valuation after first thinning. This difference is then divided by the time interval to and including first thinning giving a figure to add to each year's capital value to account for the growth in stock. In Table 15a the difference is £77 (£300–223) and the time interval 17 years. This gives a figure of £4.53 per year (£77/17) to account for capital growth. This could be taken as £4 for the first 8 years and £5 for the last 9 years.

Example 10: *The determination of the limits bounding the appropriate rate of interest*

From Table 15a we know the following values (after inflation):
Initial capital investment (I) in year 0	£161.50
Cumulative capital investment (C) in year 16	£222.76
Capital value of standing crop (S) in year 17	£300.00
The difference (D) between S and C	£77.24
The number of years (n) to 1st commercial thinning	17

(a) Upper limit percentage $= \dfrac{D}{I} \times 100 = \dfrac{77.24}{161.50} \times 100$

$= 48\%$ over 17 years
$= 2.3\%$ per year.

This final percentage may be determined using logs or with the aid of a calculator that has the function $n\sqrt{y}$.
In this example $y = 1.48$ and $n = 17$.

(b) The lower limit percentage $=$

$\dfrac{D}{C} \times 100 = \dfrac{77.24}{222.76} \times 100 = 35\%$ over 17 years

$= 1.8\%$ per year.

Therefore the appropriate rate must lie between 2.3% and 1.8%. Because most of the additional capital investment is at the end of the immature phase (Table 15, column 9) then the appropriate rate will be near to the upper rate and is, as previously stated, 2.2%.

Columns 11 to 14, Table 15a show how the capital valuation is calculated each year for the immature phase and column 15 gives the capital value from year 0 to year 16 in 1978 money values. It will be seen that when the appropriate capital growth rate is used, the value at the end of the immature phase is just less than the value at the start of the merchantable phase. Naturally the expected rate of return on the plantation over the whole rotation will be more than the above allowance because the growth rate of the stand is still accelerating towards its maximum mean annual volume increment,[8] that is when the average rate of volume increment has reached its peak. Again it is usual for the

---

[8] The point of maximum mean annual increment is not necessarily the point of maximum financial yield. This will be discussed in Part III.

unit price of logs to increase with increasing diameter and this fact will reinforce the previous statement.[9]

The above estimation of capital valuation assumed that the plantation had just reached the merchantable phase; to value a crop at any point between planting and first thinning the historic costs should be inflated to that particular point in time and then future costs estimated based on current costs and practices. Once this has been done an appropriate "capital growth rate" may be applied to the costs.

Sometimes, first and maybe second thinning are undertaken as silvicultural rather than commercial operations. In cases like these the crop is considered to be in the immature phase until the first commercial thinning and valued accordingly. Again the establishment and tending costs may be slightly more than the crop value at the first commercial thinning. Then the valuation could be a little flexible and allow the immature phase to stretch into the merchantable phase until the valuations are more or less the same. This will mean, however, that no allowance will be made for the growth of the crop. If the establishment costs are much greater than the valuation at the start of the merchantable phase then the drop in valuation should be accepted and the establishment and tending costs downgraded to give a "realistic" value to the crop in immature phase. As stated previously, if this is the case then normally the investment should not have been undertaken as its financial yield will most likely be much less than alternative investments.

Unit Area Valuation of Immature Woodlands using Current Costs and Prices

The above method of determining the capital valuation of the immature area may not be practical because extra work is involved to inflate the historic costs or detailed compartment costs have not been kept in the past. To overcome these drawbacks average current costs may be used. This eliminates the extra inflation calculation for inflation is automatically built into the costs and prices. Unfortunately, from the

[9]In sawmilling and plywood manufacture there is less waste in the primary processing stage the larger the diameter of the tree or log (also quality tends to increase with size). Therefore the purchaser of the raw material is willing to pay a higher price for the timber the larger the diameter, thus a typical price/size gradient may be established for each species/country.

TABLE 15b. *Capital*

(b) *Using present-day (1978) costs*
Land value 1978 £100 per ha.
(i) *Immature area*
Average 1978 operation costs taken from the operation analysis sheet (Table 5) or record cards (Table 7).
Compt. 7. 40 ha. 1961 1st Thinning 1978
Unit of costs and receipts are £/ha.

Year	Age	Operation S.H. nos.	Costs Direct	Costs OH	Total net cost[a]	Cumulative cost	Value at year's start
1961	0	0 to 7	145.98	24.18	170.16	170.16	0
62	1			1.33	1.33	171.49	170.16
63	2	7 & 8	4.70	2.38	7.08	178.57	174.72
64	3			1.33	1.33	179.90	185.12
65	4			1.33	1.33	181.23	189.97
66	5			1.33	1.33	182.56	194.91
67	6			1.33	1.33	183.89	199.94
68	7			1.33	1.33	185.22	205.07
69	8			1.33	1.33	186.55	210.30
70	9	10	3.50	1.50	5.00	191.55	215.62
71	10			1.33	1.33	192.88	224.72
72	11			1.33	1.33	194.21	230.32
73	12			1.33	1.33	195.52	236.03
74	13			1.33	1.33	196.87	241.84
75	14			1.33	1.33	198.20	247.77
76	15			1.33	1.33	199.53	253.80
77	16	17	25.00	5.50	30.50	230.03	259.95

(ii) *Merchantable area* (see also Tables 25, 26, 27)

Year	Age	Operation S.H. no.	Main crop after thinning Stems/ha (No.)	Main crop after thinning Volume/ha 7 cm top diam. (m³)	Main crop after thinning Value per m³ (£/ha)	Capital valuation (£/ha)
78	17	22 (1st)	1750	100	3	300
79	18		1360	115	3	345
80	19		1360	130	3	390
81	20	22 (2nd)	1360	145	4	580
82	21		1060	161	4	644

et cetera

[a]Total net cost = Gross cost less receipts (if any). Such receipts include Sale of scrub for firewood, Government grants, Tax rebates, etc.

CAPITAL VALUATION

Valuation per Unit Area

Increase in value to account for capital growth			
Increase in value during year[b]	Year's cost	Value at year's end	Capital valuation
0	170.16	170.16	170
173.39	1.33	174.72	175
178.04	7.08	185.12	185
188.64	1.33	189.97	190
193.58	1.33	194.91	195
198.61	1.33	199.94	200
203.74	1.33	205.07	205
208.97	1.33	210.30	210
214.29	1.33	215.62	216
219.72	5.00	224.72	225
228.99	1.33	230.32	230
234.70	1.33	236.03	236
240.51	1.33	241.84	242
246.44	1.33	247.77	248
252.47	1.33	253.80	254
258.62	1.33	259.95	260
264.89	30.50	295.39	295

Stems/ha (No.)	Volume/ha 7 cm top diam. (m³)	Thinning			Total capital valuation main crop plus thinning
		Value per m³ (£)	Capital valuation (£/ha)	Value of thinning removed £	
(630) 0	(20) 0	2.0	0	(40)	300
390	6	2.0	12		357
390	13	2.0	26		416
(390) 0	(20) 0	2.5	0	(50)	580
300	10	2.5	25		669
———et cetera———					

[b]Increase in value to account for real growth of capital 1.9% per year appropriate rate. Range: Upper limit 2.0%; Lower limit 1.6% (see Example 10).

valuation point of view, forest practice has not remained stationary over the years and to apply a current day establishment cost to, say, a 15-year-old plantation may seem wrong. Fifteen years ago in the United Kingdom mechanisation was in its infancy: very little ploughing and mechanical draining were carried out, hand weeding was dominant, trees were planted closer together, labour was comparatively cheap, and so on. However, the object of crop valuation is to assess the worth of the plantation with some degree of accuracy but without too much complication. Therefore, if compartment costs have not been kept, average present-day costs are the simplest, if not the best, substitute. As previously stated it also overcomes general inflation if current costs are used. However, the costs have still to be compounded up to account for the capital appreciation of the growing stock. Table 15b repeats the capital valuation as demonstrated in Table 15a only this time average current costs are used. It will be noted that in the example chosen current establishment costs are greater than inflated historic costs. Also by assumption overhead costs are constant from year to year. The cumulative capital costs amount to £230 and the appropriate capital appreciation factor is 1.9% per year.[10] The capital valuation figures are given in the final column and because average plantation figures have been used these valuation figures may be used for any immature plantation on the particular project for that specific year. Likewise the valuation of the merchantable area may be used for every stand if average figures in Table 15b(ii) are taken as average per hectare values for different plantation ages as of 1978.

Valuation of the Land

The above valuation has only taken into consideration the value of the growing stock. Figure 2 assumes that the land has a value and is represented by a constant amount over the years. The value of the land will vary from country to country and within a country from region to region. It will also vary depending, of course, on whether the area remains under forestry or is put to another use. If houses or industry are designated to a forest area then the unit value will increase considerably.

[10]Alternatively £4 could be added each year to account for the increase in capital growth, calculated as follows: £300−£230 = £70, £70/17 years = £4.12 p.y.

CAPITAL VALUATION

However, in this manual the price of forest land is taken as the price it would command in an alternative rural use, but it is assumed not to increase significantly in constant-price terms over the years and is treated as a fixed asset, which does not affect the capital value of the growing stock. Table 15b assumes its value to be £100 per hectare, giving a total value of £136,000 for the land. However, a capitalisation rent has been assumed and this is already included in the income and expenditure account. It should be adjusted from time to time to take into account an increase or decrease in land values.

Valuation of Uneven-aged Stands and Natural Forests

The valuation of uneven-aged stands or natural forests is more complicated than that of an even-aged stand but the same principles could be used. Where merchantable trees occur in an area or compartment these can be valued at the current selling price; for the immature portion of the forest an assessment will have to be made of the coverage and the various age classes represented; an average value could then be assumed per age class and this would then be multiplied by its percentage coverage to arrive at a capital value for the whole immature area.

Valuation of the Whole Crop

Once the unit area value of the different species and site classes has been determined using either actual historic costs or current costs the next stage is to value the entire growing stock by using these unit area values. A list should be made of all the compartments by area, age and species and divided into three categories: immature, mature and bare land which included compartments that have been felled during the year. Table 16 gives a capital valuation for Arbor enterprise at the end of the 1978 forest year. Average current costs and prices have been used in the valuation and on opening and closing valuation is given for each year's planting. The first section values the immature area which in this particular example accounts for 1040 hectares. It can be seen how the valuation is built up using average unit area valuation figures. The opening valuation for a particular year is calculated by multiplying the area by the previous year's unit capital value. For example, 65 ha ×

TABLE 16. *Capital Valuation Arbor Forest Enterprise Year ending 30 September 1978*
Average current costs and prices assumed

P. Year	Age (years)	Compts	Species	Unit C.V. (£/ha)	Area (ha)	Opening (£)	Closing (£)	Increase/Decrease(—) (£)
				Immature phase				
1978	0	2 & 10	S.S. (J.L.)	170	260.5	0	44,285	44,285
1977	1	15	N.S.	174	65	11,050	11,310	260
				—et cetera—				
1963	15	24	S.S.	260	20	5080	5200	120
1962	16	23	S.S.	295	10	2600	2950	350
Sub-total					1040	160,492	209,730	49,238
				Merchantable phase				
1961	17	22	S.S.	300	10	2950	3000	50
1960	18	21	S.S.	357	10	3000	3570	570
				—et cetera—				
1924	52	12	S.P.	2400	2	4600	4800	200
Sub-total					318	1,008,652	1,085,094	76,442
				Final felling				
1925	53	11	S.P.	2400	2	4800	0	—4800
Total capital value					1360	1,173,944	1,294,824	120,880

£170 giving £11,050 opening valuation for the 1977 planting. Likewise, the closing valuation is calculated by multiplying the area by the year's unit capital valuation—65 ha × £174—giving £11,310 closing valuation. The difference between the closing and opening valuation measures the increase (or decrease) in the forest's valuation. It will be seen in the example (Table 16) that the valuation of the immature area of 1040 ha in 1978 came to £209,730 and showed an increase of £49,238 over the previous year.

The second section values the mature area using current prices and actual or management table volumes. The Unit Capital value is obtained from Table 15b(ii). In the example 318 ha are valued at £1,085,094, an increase of £76,442. It will be noticed that although the merchantable area only occupies 30% of the land it has a value five times greater than the immature area. This is to be expected, for the value of the trees once they have reached the commercial thinning stage accelerates rapidly.

The last section values the area felled in that particular year using actual figures, both volume and value. These figures may differ slightly from the closing valuation of the previous year but if the valuation has been kept up to date the difference will not be significant. Of course, when the trees are felled the capital has been liquidated and therefore the money obtained from the sale has to be subtracted from the capital value; this is also true for the realisable value of the thinnings, but the income is recorded in the income and expenditure account.

The valuation of the growing stock indicates how much capital is tied up in the immature and merchantable stages, and illustrates the high ratio of growing stock value to realisable income from thinning and felling. In the particular example, this ratio is about 5:1 but in a normal forest the ratio could be as high as 35:1. The capital valuation also indicates whether the forest is increasing or decreasing in value from year to year. Once this is known a profit and loss statement may be compiled for the year. First, however, other methods of capital valuation will be discussed.

The Realisation Capital Value

The realisation capital value is very similar to the previous method except that it is assumed that the forest in the immature stage does not have any realisable value; the realisation value neglects the value of the forest in the immature stage and only values the merchantable stage volume using current market prices. In the example shown, Table 16, the realisation value is £1,085,094, some £209,730 less than the total capital value. In a normal forest neglecting the value of the immature stage will only mean about a 3% drop in value as compared to the first method. However, if there is a large proportion of the area in the immature stage then the capital value will be significantly less than in the first method. Even though the immature area does not have a realisable timber value, capital has been invested in it and it has a market value because of its potential. Therefore, to neglect it is to deny that it has a value which is not the case, thus although this method is easier to apply than the "actual value" method, the under-estimation of capital value is a major drawback.

The Expected Yield Capital Value
(a) *Potential capital value*

As will be discussed in Part III of this book, the financial yield for a forest enterprise, or a single plantation, may be determined by making various assumptions about costs and prices. The financial yield is a rate of interest that is earned on invested capital over the life of a project or plantation. That is to say, when all plantation costs and revenues are discounted to the year of planting at a specific rate of interest discounted costs will be equal to discounted revenues. The specific rate of interest may be obtained by applying the general formula, Formula 3.

Formula 3: Financial yield rate of interest

$$\frac{I_n - E_n}{(1.0p)^n} + \frac{I_{n-1} - E_{n-1}}{(1.0p)^{n-1}} + \frac{I_{n-2} - E_{n-2}}{(1.0p)^{n-2}} \ldots I_0 - E_0 = 0$$

$$\text{or} \quad \sum_{0}^{n} \frac{(I_x - E_x)}{(1.0p)^x} = 0,$$

where *I* equals the income in any particular year n, $n-1$; $n-2$ to the year of planting (0),

E equals the expenditure in any particular year,

n equals the rotation age of the plantation,

p is the rate of interest in percent,

x is any year between n and zero.

As will be explained in Part III, the financial yield is determined by applying various rates of interest in the above formuale, until the unique rate is found. Once the financial yield has been determined this rate of interest can be applied to net invested capital, that is the expenditure minus the income, on any plantation, at any age and a potential capital value will be obtained. This potential value is higher than the actual value except in year zero and immediately before final felling because to fell at any point before the rotation age of maximum financial yield[11] would mean that money is being sacrificed and the return on investment will be lower than if the crop had been allowed to grow to financial

---

[11] If the rotation age is not determined according to financial yield but using other criteria such as maximum volume production, then it is likely that the potential value will, for some time, be below the actual value. The cross-over usually occurs towards the end of the rotation. In cases like these, the actual value should be assumed after the cross-over.

maturity. The determination of capital value using the potential capital value method may be illustrated by a simple example, Example 11. Assume that the costs and revenues on a 1-ha Christmas-tree plantation on a 6-year rotation are as follows:

Example 11a: *Costs and revenues on a 1-ha Christmas-tree plantation*

Units £ in constant money values

Year	Costs	Revenues	Net costs (net invested capital)
0	170		170
1	40		40
2	40		40
3	40	35	5
4	5	65	—60
5	5	135	—130
6	5	220	—215
Total	305	455	—150

The negative values in the last column imply that capital is being liquidated and turned into income.

Formula 3 (p. 112) has been applied to the above costs and revenues and it is found that the financial yield is 10%. That is at this rate of interest the discounted revenue (£279) equals the discounted costs (£279). This interest rate is then applied annually to the net invested capital to determine the potential capital value as follows:

Example 11b: *Potential capital value each year on a 1-ha Christmas-tree plantation*

Units £ in constant money values

Year	Capital value (C.V.) at the start of the year	Increase in C.V. due to growth potential of the crop (10% p.y.) multiplier 1.10	Value added during the year (net invested capital)	C.V. at the end of the year
0	0	0	170	170
1	170	187	40	227
2	227	250	40	290
3	290	319	5	324
4	324	356	—60	296
5	296	326	—130	196
6	196	215	—215	0

Naturally the capital value of the crop starts at zero before planting, and finishes at zero when it has been felled and sold.

The potential capital value of the 1-ha Sitka spruce plantation (Table 15) at different ages on a 4- or 35-year rotation has been worked out (Appendix II) using the costs as given in Table 15b and the revenues as given in Table 27. The maximum financial yield of this crop with the above costs and revenues is 10.4% and the rotation age for this financial yield is between 34 and 35 years. This potential capital value may then be compared to the actual value as determined on Table 15b. Appendix II of this book gives a comparison between actual capital value and potential capital value using the costs and revenues described above. It will be observed that the potential capital value except in the first and the final year is considerably greater than the actual capital value. This method of valuation is more favourable to the owner/seller for it considers future potential of the crop but is more subjective than the actual value method. With the potential valuation method financial yield has to be determined and this is dependent on the assumptions that are made about future costs and revenues whereas with the actual valuation method, the crop is valued at what it would fetch if it were liquidated today. Normally, of course, plantations would not be felled until the forest manager considers that they have reached financial maturity, that is when their potential has been fulfilled, and this is what the potential valuation method tries to assess.

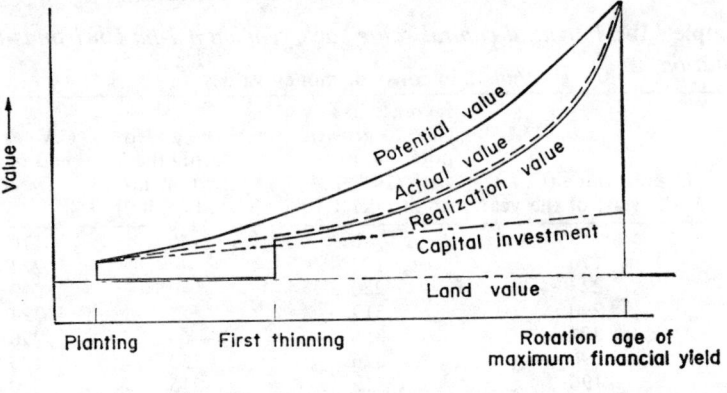

Fig. 3. Capital valuation of a normal forest area or a single-aged crop at different ages using various assumptions.

The difference between the two valuations is illustrated in Fig. 3. The potential-values curve increases at a uniform percentage rate each year whereas for the actual value curve, the percentage rate varies from year to year depending on the growth rate of the crop and the diameter/price relationship for roundtimber; usually it starts off with a very low rate of growth, and gradually increases to a maximum. If a forest area was sold before it reached financial maturity the actual market price should lie somewhere between these two values, although it is possible that it is even above the potential value if that particular area was specifically wanted for a particular purpose. However, the market price should not fall below the actual value but, of course, this depends on whether the forester is fully aware of market prices and has undertaken a detailed capital valuation.

When using the potential capital value method, assumptions have to be made about future costs and prices for these values are an integral part of the calculation. It is usual to assume that costs and prices will remain unchanged and then work out the financial yield accordingly. No such assumptions have to be made when using the actual valuation method because the valuation refers to the present day and is based on present-day costs and prices, therefore it has fewer imponderables. Both methods may either use historic or current costs and prices for operations that have already occurred, but as stated previously if historic values are used then account should be taken of inflation. The potential value method takes into account the fact that even though the stock has an actual value this value is lower than its worth, otherwise the manager would recommend the area be felled rather than left to grow. When assessing an area for insurance, damage, or appropriation then this method is more beneficial to the owner; of course, for taxation and book-keeping purposes the actual-value method appears to be more suitable and from an accounting viewpoint more correct. However, the important point to note is not which method is right but that a capital valuation should be carried out in order to make the forest manager aware of the value of the asset he is managing. Either of the above two methods will give approximately the same results for the increase or decrease in capital value and therefore the profit and loss account will be similar whichever method is chosen.

(b) *The expectation value*

A variation on the potential-value method is the expectation value; whereas the potential yield takes all costs and prices and determines a unique rate of interest to apply to the valuation, the expectation value is only concerned with future income and expenditures and these are discounted to the present day at a rate of interest that is mutually agreed between two parties—the owner and, for example, the insurance company or a compulsory purchase body. This method has long been used in Scandinavia especially when land is being assessed for appropriation, sale or insurance purposes.

The expectation value of a single stand at the age of m years may be calculated using the following formula, Formula 4.

Formula 4: Expectation value

$$EV_{am} = \frac{I_n - E_n}{(1.0p)^{n-m}} + \frac{I_{n-1} - E_{n-1}}{(1.0p)^{n-1-m}} \cdots \frac{I_{m+1} - E_{m+1}}{(1.0p)^{m+1-m}}$$

or

$$EV_m = \sum_{m+1}^{n} \frac{(I-E)}{(1.0p^{x-m})},$$

where I is the anticipated income in any particular year from age n down to the year $m+1$,
E equals the expenditure in any particular year,
n equals the rotation age of the stand,
m equals the age of the stand today,
p is the chosen rate of interest,
EV_m is the expectation value at the *end* of year m,
x is any date between n and $m+1$.

It should be emphasised that both I and E only apply to *future* incomes and expenditures. If, for example, the stand is 5 years old then no establishment costs are included. Also the expectation value (EV_m) is calculated at the end of the particular year (m) so no income or expenditure for that year is included in the calculation because these sums of money have already been spent or collected. The calculation of the expectation value is just like that for the potential value only in reverse. Instead of starting from year zero and compounding up the net invested capital, the net income, that is income minus expenditure, starting from

CAPITAL VALUATION

the final felling, is discounted down. If a rate of interest precisely the same as the potential value has been chosen then the resulting capital value for any particular year will exactly match that of the potential value. However, if the chosen rate of interest is lower, then the expectation capital value will be higher than the potential capital value and vice versa for a higher interest rate. How is the rate of interest chosen? In theory it should be chosen by undertaking financial yield calculations on each plantation and indeed this forms the basis for fixing the rate, but in practice the rate is fixed by compromise between the owner of the forest or his representatives and the insurance or expropriation body; the owner favours a low interest rate, whereas the other party argues for a high rate. In countries where this method is applied disputes about the interest rate go to arbitration. Books of tables are published by forestry bodies giving guidelines to rates under various conditions and these form a basis for compromise and arbitration. The expectation-value method is usually used for immature plantations, whereas the realisation value is the one adopted for mature stand. The expectation value may be illustrated by applying the costs and revenues already used in Example 11. This is shown in the following example, Example 12.

Example 12a: *Costs and revenues on a 1-ha Christmas-tree plantation*

Units £ in constant money values

Year	Costs	Revenues	Net revenues
0	170		−170
1	40		−40
2	40		−40
3	40	35	−5
4	5	65	60
5	5	135	130
6	5	220	215
Total	305	455	150

The expectation value for any one year only considers future net revenues (revenues minus costs). Therefore the capital value at the end of year 6 equals future net revenue from this plantation divided by $1.0p$, where p is the chosen discount rate in percent (for rates of 10% or more the division becomes $1.p$). If we assume the chosen discount rate equals 10% then;

(a) Future net revenues at the end of year 6 = 0.
Therefore C.V. end year 6 = 0/1.10 = 0.
(b) Future net revenues at the end of year 5 = £215.
Therefore C.V. end year 5 = 215/1.10 = £195.
(c) Future net revenues at the end of year 4 = 215 (year 6)+130 (year 5).

Therefore C.V. end year 4 = $215/1.10^2 + 130/1.10 = \frac{195+130}{1.10} = 296$

and so on.

The annual expectation capital value of the above plantation using three rates of discount, namely 9, 10 and 11%, is shown in Example 12b.

Example 12b: *Expectation capital value using varying discount rates, on a 1-ha Christmas-tree plantation*

Units £ in constant money values

Year	Net revenue	Capital values at the end of the year Discount rate %		
		9	10	11
0	−170	181	170	159
1	−40	237	227	217
2	−40	299	290	281
3	−5	330	324	317
4	60	300	296	292
5	130	197	195	194
6	215	0	0	0

When a discount rate lower than the financial yield is used then the capital value is higher than the potential value and vice versa. This is why a seller favours a "low" discount rate and a buyer a "high" discount rate.

If the discount rate is much below the financial yield percentage then the capital valuation of the crop in year 0 will be substantially above the initial establishment costs. This implies either that the forest manager is efficient for his costs are below the industry's average, or the crop is overvalued or there is no free entry into the forest sector and the owners of the forest areas are in a monopolistic cartel, otherwise a person would be able to buy land and plant a crop for prices not greatly different from the average costs for the industry. If on the other hand the chosen discount rate is much above the financial yield percentage then the opportunity cost for capital must be much greater in other industries, or the forest manager is inefficient and has high unit costs.

It will be seen that this method of valuation is more subjective than others for it tries to evaluate the intrinsic value of an area, a value that no two people will agree upon.

Valuation assuming a "Normal"[12] Forest—the Capitalisation Value

Another method of determining the potential capital value of a forest is to assume that it has a uniform age class distribution. If the forest is "normal" then the capitalisation of a permanent annual income is given by Formula 5.

Formula 5. Capitalisation value
$V_o = R/0.0p$ where V_o = capital value,
R = average net annual income,
p = the expected financial yield in per cent.

For example, the net annual income of a normal forest on a 40-year rotation may be of the order of £120 per hectare or for 1360 hectares £163,200. If the anticipated financial yield is 9%, then the capital value of the forest is £1.81 million (£163,200 ÷ 0.09) or £1333 per hectare.

The forest manager will have to estimate the rotation and average yield, then apply current costs and prices to determine the average net annual income. The rate of return on the investment or the financial yield is then worked out by methods described in Chapter 11 and so the capital value can be estimated. Of course, in a "normal" forest there will be no increase or decrease in the capital valuation except through a change in costs and prices and so the income and expenditure account will be the same as the profit and loss account.

The capitalisation value should only be used when the rotation age is fixed round about the age of maximum financial yield. At any other age, the capitalisation value will be considerably below the actual value if the financial yield percentage is used, thus underestimating the forest's value. Bearing in mind this drawback, the capitalisation value is a quick method of determining the potential value of a forest.

[12] Normal forest: equal distribution of all age classes and equal volume yield per year/period. See p. 90.

Accountant's Capital Valuation

From the accountant's point of view, the capital valuation as described above is not correct. Unlike most capital, the forest is living capital and has the power of growth. The above valuation tries to take this fact into account, hence a percentage increment has been added to the crop value each year. A strict accounting method should only value the crop according to the actual net capital expenditure spent on the enterprise, no allowance being made for growth. In theory it should be applied both to the woodland in the immature and merchantable phases, care being taken to exclude such revenue expenditure as thinning and felling costs. The value of the woodlands using this method is always much less than the previous method and may be favoured when valuing a forest liable for tax or death duty, but not when selling or insuring the woodlands.

Choice of Method

Several methods of capital valuation have been described and arguments can be made to justify each method. It will be seen that because valuation is based on uncertainties about prices, costs, areas, ages, volumes, etc., it is not an exact undertaking. Nevertheless a capital valuation is a useful and necessary operation. It gives an estimate of the value of the growing stock, and indicates whether the capital investment is increasing, decreasing or remaining more or less constant. This latter piece of information may be used for compiling a profit and loss account and the "actual-value" method is as accurate as any of the other methods described. However, in the event of selling or insuring the forest the potential value should be used as a measure.

Chapter 10

PROFIT AND LOSS ACCOUNT AND BALANCE-SHEET

Introduction

The previous two chapters have described how an income and expenditure account is compiled and capital valuation undertaken. All that remains to be done is to draw up a profit and loss account and a balance-sheet to complete the financial statement.

Profit and Loss Account (P/L Account)

The P/L account is exactly the same as the income and expenditure account (Table 12), except that the valuation increase/decrease has been included on the credit side of the account, and the balancing item of net profit is shown on the debit side. This is illustrated in Table 17.

TABLE 17. *Profit and Loss Account*

P/L account for Arbor enterprise (forest section) for the forest year ending 30th September 1978

Debit	£	Credit	£
To Management supervision and labour	10,393	By Sales of timber	50,052
To Vehicles and machines	12,073	By Sales of other products	39
To Capital expenditure depreciation	0	By Grants, etc.	0
To Consumable stores	7077	By Miscellaneous receipts	140
To Buildings and land	2233	By Valuation increase/ decrease	120,880
To Administrative expenses	220		
To Profit (before taxation)	139,115		
	£171,111		£171,111

If the enterprise is liable for taxation this can be deducted from the above profits and the profits after taxation shown separately.

Balance-sheet

In order to complete the financial accounts a *Balance-sheet* for the forest enterprise should be given. This merely sets out to show the opening and closing valuation of the forest and other capital assets belonging to the enterprise, as well as the profit (or loss) made on the year's activities. It has been assumed in the illustrated example that the forest enterprise does not own the land or buildings on the land but pays a rent for the use of the above. If the forest enterprise did own the buildings and land then capital accounts similar to the "Machine Capital Account", Table 13, would be drawn up for the enterprise. The only items of capital equipment that the forest enterprise owns are the machines, and a capital account has been described previously for this (Table 13). Therefore, a balance-sheet can be compiled using the information already obtained.

TABLE 18. *Balance-sheet, Arbor Enterprise (Forestry Section) Year ending 30th September 1978*[1]

Previous year £	Assets Current Assets	£	Previous year £	Liabilities Investment	£
11,000	Balance at bank	29,235	1,067,503	Capital plus interest	1,185,653
	Fixed Assets			Net profit for	
	Equipment and machinery value at start of year £998 less depre-		118,439	the year	139,115
998	ciation £289	709			
	Growing Stock Forest closing				
1,173,944	valuation	1,294,824			
1,185,942		1,324,768	1,185,942		1,324,768

On the *assets side* the balance at the bank is the current year's surplus added to that of the previous year's, plus interest (if any). The fixed

[1] By tradition, assets in the U.K. are placed on the right and liabilities on the left, but in most other countries they are more logically the other way round.

assets consist of the written-down values of plant, machinery, buildings, land, etc., and the last item is the growing-stock valuation. The liabilities simply consist of the net profit for the year (after tax) and the Investment capital plus interest, which is just a balancing item representing the Capital investment of the enterprise.

The net profit in Table 18 gives a figure for the year's activities but because the forest is not "normal", the potential profit may be greater or smaller than the actual profit. It is very unlikely, and may not be desirable, that the forest will ever be normal. In order to overcome this difficulty the *potential return* per stand/per forest can be calculated and this is demonstrated in the next chapter.

Part III
THE FINANCIAL YIELD

Chapter 11

DISCOUNTED EXPENDITURE

Introduction

The previous chapter demonstrated how a profit and loss account is built up for the forest enterprise for any one year. But such an account cannot give a true picture of the financial yield from the enterprise unless the forest is normal or at a first approximation near normal. Now it may be that normality is a state never to be realised because (a) the forest owner or woodland manager does not believe in the concept of "normality", (b) natural hazards upset the plan, or (c) future demand conditions will constantly change and make it financially inexpedient to pursue such a concept. But the manager or owner should want to know what kind of return to expect on the forest investment without having to wait until a "normal" state is reached. He should also want to know what each plantation will yield. This can be done if certain "heroic" assumptions are made, the principal one being about the future stumpage price of timber.

The financial return for each even-aged plantation may be worked out by using variations of the compound interest formula; either the incomes and expenditure per plantation can be compounded up to the end of the rotation, or they can be discounted back to the start. The latter method is the one recommended in this part because it gives the result in a present-day concept. It shows how important the establishment costs are in determining financial yield and the measure by which returns, especially early returns, contribute to profitability; secondly, by adopting a discounting process one can compare and contrast different rotation lengths by assuming each crop is repeated in perpetuity, i.e. discounting back from infinity. This enables one not only to find the maximum financial yield but also to determine the optimum financial rotation length, by definition the rotation of maximum financial yield.

However, the discounting process only tells the manager the expected yield. Once a crop has been felled the actual yield can be found, if information on costs and revenues are available, by compounding the various items to the present day. Again, assuming, say, a second thinning has just taken place, the manager could work out the expected yield by combining the discounting and compounding processes. Likewise, he could use the formulae to see what percentage increase in crop yield is necessary in order to cover the cost of fertiliser application, or the extra money required in order to justify a pruning.

The process of compounding and discounting is time-consuming but worth while. However, calculators and computers can ease the burden. It is a tool which, once mastered, will bring home the importance of choice of species, establishment costs and time, on forest profitability.

Why is the concept of compound interest—a method of expressing present-day sums in terms of future values (or vice versa)—brought into forestry? The main reason, as has been stressed in this book, is that forestry is a business and like any other business it expects to earn a return on invested capital, be it in plantations or natural forests; by using the compound-interest formula the return can be expressed in simple terms. Again if governments or individuals want to compare investments either within the forestry sector or between different sectors, they must have some common criteria and percentage money yield or return is an accepted one. Basically varying rates of interest are applied to all costs and revenues and these are discounted back to give a series of values in present-day terms. The calculated return is then that rate of interest which equalises discounted costs and revenues.

This part sets out in detail the method of determining the financial yield illustrating the methodology with a plantation crop of *Picea sitchensis* (Sitka spruce) with an average yield of 18 m^3 per hectare (257 ft^3 or 202 H.ft^3/acre).[1]

Discounting Process—Age Assumptions

In this manual the age of a plantation is reckoned in the same way as the age of human beings, that is it is not 1 year old until the first anni-

[1] H.ft^3 = hoppus cubic foot = 1.273 true cubic feet.

versary, etc. All costs/revenues in the first year (Year 0) are not discounted but taken at their full value. It is only after the plantation is 1 year old that the discount process applies. This in theory assumes that all costs in year 0 occur at the start of the year, whereas in fact they are spread out throughout the year; to be strictly accurate the discounting should be in days (or weeks) rather than years. Some publications assume that all costs in the first year occur at the end of the year, that is on the first anniversary, and therefore they discount all first-year costs by the 1-year discount factor and so on. These publications also assume that the forest crop is cut exactly on its anniversary so that a crop on a 40-year rotation is cut on the 40th "birthday". This manual assumes more logically that trees can be cut any time within the year; a crop on a 40-year rotation may be cut on its "birthday" or any time during its 41st year up to but not including the 41st "birthday". Differences, therefore, will occur between the two assumptions. The student or manager may use either method, but once one method has been chosen it must be adhered to. If comparisons are to be made with other projects whether forestry or between forestry and other potential investments, then it is again important that uniform assumptions are made.

Financial Yield—Cost Assumptions

The assumptions made in order to determine financial yield can be stated as follows: Costs are *constant* over the years; no allowance has been made for inflation or productivity increases. All costs are therefore expressed in current costs and in present-day money value. Average costs have been used. The specific costs include ascribable overheads as well as direct costs but do not include "unit area" overheads which are given separately. These costs are shown in Table 19 and have been extracted from Tables 7b and 15b.

It is assumed that the gross costs do not vary much from plantation to plantation and therefore the average is a reasonable cost to use for the entire forest estate. However, if the manager knows that the costs on individual areas vary greatly from the average, or the costs vary from species to species, then these costs should be substituted and used in the discounted expenditure process for the particular plantations.

TABLE 19. *Average per Hectare "Capital" Expenditure*

Year	Type of expenditure[a]	Gross cost	Income grants[b] tax rebate	Net cost (units—£)
0	Weeding and initial establishment	168.83		168.83
2	Weeding and beating-up	5.75		5.75
9	Early cleaning	3.67		3.67
16	Brashing	29.17		29.17
0–16	Per hectare overheads per year	1.33		1.33

[a]The costs given in Tables 7b and 15 include the constant overhead cost of £1.33 p.y. This cost has been subtracted from the operation costs.

[b]The income, grants and tax rebates, if any, should be credited to the years in question and deducted from the gross cost to give a net cost. It is the net cost that is used for financial yield calculations.

It has been assumed that no income is obtained either from sale of scrub, grants or tax rebate. If income is obtained this should be deducted from gross costs.

It will be noticed that the expenditure is given up to year 16—the year before first thinning—and this is termed "capital" expenditure. From the first thinning onwards it is assumed that the revenue is sufficient to cover the costs of the various operations, *including the unit area overhead costs*; therefore, the revenue figures given in Table 27 are net of direct and overhead expenditure—"revenue" expenditure.

Discounted Expenditure (Constant Costs) Single Rotation

The process of discounting is very straightforward; it is the compound interest formula in reverse; these formulae are as follows (Formulae 6a and 6b):

Formula 6a: Compound interest formula

Compound interest $V_n = V_o(1.0p)^n$,

V_n = value at year n,
V_o = present-day value,
p = rate of interest in per cent.

Formula 6b: Discount interest formula

Discount formula $V_o = V_n \times \dfrac{1}{(1.0p)^n}$.

The value of $\dfrac{1}{(1.0p)^n}$ is given in tables of compound interest for different years (n) and different rates of interest (p). If the manager wishes to work out values and does not have compound-interest tables then he can use logarithmic tables or an electric calculator. The discounted value of a yearly sum can also be worked out by a variation of the above formula, Formula 7, and is:

Formula 7: Discount value of a constant yearly sum

$$V_o = \frac{R(1.0p)^n - 1}{0.0p \times 1.0p^n},$$

R = constant yearly sum,
n = length of expenditure period (rotation).

The above multiplier is usually given in the compound interest tables. However, this formula only discounts a constant sum back to year 1 and overheads occur in year 0 as well. Therefore, in order to account for a constant expense (or income) commencing in year 0, ONE must be added to the multiplier. The formula now becomes

$$V_o = R\left(1 + \frac{1.0p^n - 1}{0.0p \times 1.0p^n}\right).$$

It is this formula that is used for constant annual "unit area" overheads. If these overheads vary considerably from year to year then it would be misleading to use the above formula and each year's overhead cost should be discounted back separately.

Table 20 gives multipliers used to calculate discounted expenditure. A wide range of interest rates have been chosen in order to accommodate

TABLE 20. *Multipliers used to calculate Discounted Expenditure*

Age (years)	Rates of interest chosen (%)			
	5	7	10	12
0	1.000	1.000	1.000	1.000
2	0.907	0.874	0.826	0.797
9	0.645	0.544	0.424	0.361
16	0.458	0.339	0.218	0.163
0–16	11.838	10.446	8.823	7.974

the various variable assumptions. The woodland manager need not take such a range.

All that has to be done once the rates of interest are chosen is to multiply the figures in Table 20 by those in Table 19. For example, £5.75 (age 2 years) multiplied by 0.907 $\left(5\% \text{ discount multiplier } \dfrac{1}{(1.0p)^n}\right)$ gives £5.22. That is, the weeding and beating-up cost of £5.75 2 years hence discounted back at 5% is valued at £5.22 today. In other words, if one could invest money at 5% one would have to invest £5.22 today in order to pay for a weeding and beating-up on a hectare 2 years hence. If the rate of interest had been 7% the discounted revenue would be £5.03 and so on. Similarly, the brashing cost of £29.17 (age 16 years) is multiplied by 0.458 (5%) to give a discounted value of £13.36. Table 21 gives the average per hectare discounted expenditure.

TABLE 21. *Discounted Expenditure per Hectare (assuming single rotation and constant costs)*

Year	Expenditure (£)	Rates of interest (%)				Alternative expenditure[b]	
		5	7	10	12	1	2
0	168.83	168.83	168.83	168.83	168.83	338	110
2	5.75	5.22	5.03	4.75	4.58	11	9
9	3.67	2.37	2.00	1.56	1.32	7	5
16	29.17	13.36	9.89	6.36	4.75	58	19
0–16[a]	(per yr) 1.33	15.74	13.89	11.73	10.61	3	2
(Total)	230.03	205.52	199.64	193.23	190.09	465	177

[a]Overhead costs of £1.33 per year = 17 × 1.33 (0–16 = 17 yr) = £22.61.
[b]Current costs have been assumed. However, two alternative costs 1 and 2 are given to illustrate a number of points which are discussed later. One alternative is approximately double and the other approximately 75% of the current costs.

An examination of Table 21 reveals that the establishment costs (0–4 years) account for about 80% of discounted expenditure. If savings are to be made, this is the area of costs that should be attacked with most vigour. Because this particular expenditure occurs at the beginning of the period, greater proportionate savings will be made if costs can be reduced or deferred. £10.00 expenditure deferred from year 0 to year 9 will save £5.76 at 10% interest (£10.00 (year 0) − £10.00 × 0.424 (year 9)).

DISCOUNTED EXPENDITURE

As the discounted rate of interest is increased from 5 to 12%, the reduction in discounted expenditure is only relatively small when compared with the reduction in discounted revenue. This is because the bulk of the expenditure occurs at the start of the rotation whereas substantial income does not usually start to flow in for at least 20 years. This will be seen by comparing Table 21 with Table 29.

This formula is used to find out the financial return of a crop and to compare the returns from different crops over the same rotation.

Discounted Expenditure—Infinite Rotations

The costs given in Table 21 are the discounted expenditure for a *single rotation* only. Strictly speaking, in order to compare crops with different rotation lengths, similar time periods should be used. The discounted costs and revenues of three 40-year rotations should be compared with those of two 60-year rotations. However, the costs and revenues of the first rotation bulk very large in the final result, especially so the higher the rate of interest and the longer the rotation. This fact is illustrated in Table 22.

TABLE 22. *The Effect of Rotation Length and Discount Rate on Expenditure. Assume £10 is spent now and in n years' time. Values discounted to year 0*

Rate of interest	$n = 0$	10	20	30	40	50	60	70	80
5%	10.0	6.1	3.8	2.3	1.4	0.8	0.5	0.3	0.2
7%	10.0	5.1	2.6	1.3	0.7	0.3	0.2	0.1	0.0
12%	10.0	3.2	1.0	0.3	0.1	0.0			

A recurring cost of £10 at 12% on a 40-year rotation will only increase the final cost by 1%, after one rotation and further additions from future rotations may be neglected. Nevertheless, the costs (and revenues) have been discounted back from infinity in order to compare different rotation lengths and to see if one-rotation calculations can justifiably be used as a first approximation.

At a particular rate of interest let the discounted expenditure (and income) for one rotation equal Vn, and let the expenditure (and income) in future rotations be constant and equal to Vn: then we want to know the sum of:

134 COST AND FINANCIAL ACCOUNTING IN FORESTRY

$$\begin{array}{cccc} \text{Year:} & 0 & n & 2n & \ldots & Zn \\ & Vn & +Vn & +Vn & \ldots & +Vn \end{array}$$

discounted back to year 0, Formula 8. This is equal to:

Formula 8: Infinity formula

$$Vn + \frac{Vn}{(1.0p)^n} + \frac{Vn}{(1.0p)^{2n}} \ldots + \frac{Vn}{(1.0p)^{Zn}} =$$

$$Vn + \frac{Vn}{(1.0p)^n - 1} \quad \text{or} \quad Vn\left(1 + \frac{1}{(1.0p)^n - 1}\right).$$

Table 23 gives values of

$$\left(1 + \frac{1}{(1.0p)^n - 1}\right)$$

for different rotation lengths. By subtracting ONE from the multiplier in this table the part played by future rotations to costs and receipts is revealed, i.e. the value

$$\left(\frac{1}{(1.0p)^n - 1}\right).$$

For example, at 5% and 20 years' intervals future rotations increase the cost (revenue) by 60%, whereas at 12% and 40-year intervals they only add 1% to the cost (revenue).

TABLE 23. *Multiplier used to discount back from Infinity*

Values of: $1 + \dfrac{1}{(1.0p)^n - 1}$

Rotation length (n) years	Rates of interest (%)			
	5	7	10	12
6	3.94	3.00	2.30	2.03
10	2.59	2.03	1.63	1.47
17	1.77	1.46	1.25	1.17
20	1.60	1.35	1.17	1.10
23	1.48	1.27	1.13	1.08
27	1.37	1.19	1.08	1.05
31	1.28	1.14	1.06	1.03
35	1.22	1.10	1.04	1.02
40	1.17	1.07	1.02	1.01
45	1.13	1.05	1.01	1.01
50	1.09	1.04	1.01	1.00
55	1.07	1.02	1.01	1.00
60	1.06	1.02	1.00	1.00
70	1.03	1.01	1.00	1.00

Not all compound-interest tables give this multiplier factor. However, it may be worked out by using a multiplier already given, namely $(1.0p)^n$. If n = rotation age and p = the rate of interest in per cent, then look up the particular value of $(1.0p)^n$. Subtract ONE from the value and divide the answer into ONE to give the multiplier for the part played by future rotations in discounted costs and revenues. To this result add ONE to account for the current rotation. This is illustrated in Example 13.

Example 13: *The calculation of the infinity multiplier*

The discounted cost of a plantation on a 25-year rotation is £206 at 5% discount rate. What will be the discounted cost if this rotation is repeated indefinitely, assuming constant costs?

$$\text{Rotation } (n) = \text{years}$$
$$\text{Discount rate } (p) = 5\%$$

From compound interest tables the value of $(1.05)^{25}$ = 3.39
Future rotations multiplier = $1 \div (3.39 - 1)$ = 0.42
Infinity multiplier = $1 + 0.42$ = 1.42
Discounted "infinity cost" = £206 × 1.42 = £293.

That is the discounted cost of £206 for one rotation of 25 years at 5% repeated indefinitely amounts to £293. Similarly at 10% discount, £193 (Table 21) on a 25-year rotation gives an "infinite value" of £213.

In place of a single-cost curve, as is the case for one rotation, a series of curves can now be drawn, one for every different rotation length (discounted from infinity). Table 24 gives the values used in plotting these curves; it is evaluated by combining Table 21 with Table 23. For example, a gross cost of £206 (5%) is multiplied by 1.60 to give a discounted expenditure of £330 (at 5% and 20-year periods). To put it another way, a sum of £328 would have to be found now and invested at 5% in order to pay off a debt of £206 at once and again every 20 years. Likewise, a sum of £190.58 is needed to pay a debt of £190.01 now and every 50 years if it could be invested at 12%.

It can be seen from Table 24 that the greater the interest rate and the longer the rotation, the nearer the discounted expenditure approaches the single-rotation costs.

It could be argued that by using constant costs, rotation after rotation,

TABLE 24. *Value of Discounted Expenditure, discounted back from Infinity with Various Rotations and Interest Rates*

Rotation length (n) years	"Infinity" discounted expenditure (£) Rate of interest (%)			
	5	7	10	12
20	330	269	226	209
31	264	227	205	196
35	251	219	201	194
40	241	213	197	192
45	232	209	195	192
50	225	207	195	190
Single-crop discounted expenditure				
	206	200	193	190

an unrealistic assumption is being made. On the one hand, the capital cost of establishing a plantation should be greater than the cost of replanting in subsequent periods—draining and ploughing may be unnecessary and so reduce costs. On the other hand, increasing labour costs, if not offset by increased productivity—greater manual efficiency or the substitution of machines—will tend to increase costs. The reduction may balance the increase, but it does not matter to a great extent unless the difference is excessive because future costs (and revenues) usually contribute but a small amount to the final discounted total (see Table 23). A reduction in the initial establishment cost of 50% from £170 per hectare on the first rotation of say 31 years (total capital cost £230—see Table 21) to one of £85 in subsequent rotations at 5% interest reduces the discounted expenditure from £262 to £240 (that is by 10%). At 10% interest rate, with the same rotation, the reduction is from £205 to £200 (just 2½%). Therefore, the assumption of constant costs (and revenues) does not unduly upset the calculations.

Chapter 12

DISCOUNTED INCOME

Price Assumptions

As with costs, there are certain assumptions to be made in order to determine the financial yield. Prices are *constant* over the years. No allowance has been made for general inflation or a real increase in timber prices. All prices are therefore expressed in current prices and in present-day money value.

The prices given in Table 25 are net of operations costs which are taken to include ascribable overheads as well as unit area overheads. If species difference is neglected, the stumpage price of timber may vary because of two factors: the larger the tree the more valuable it becomes per unit volume; the larger the tree the cheaper it becomes per unit volume to fell and extract. This has been taken into consideration when assessing the unit price of roundwood. This is more or less synonymous with the position apertaining today. The price/size relationship is illustrated in Table 25.

TABLE 25. *Price/Size Relationship Standing (or Net Rideside) Price*

Vol/stem (m³)	Price per m³ (£)	
	Thinning	Felling
< 0.05	2.0	2.5
0.05 — 0.10	2.5	3.0
0.10 — 0.15	3.5	4.0
0.15 — 0.25	4.5	5.0
0.25 — 0.35	6.0	6.5
0.35 — 0.55	8.0	8.5
0.55 — 0.80	10.0	10.5
0.80 — 1.20	12.0	12.5
1.20 — 1.60	14.0	14.5
> 1.60	16.0	16.5

Using the above price/size gradient, the net value of the thinnings and fellings from *Picea sitchensis* (Sitka spruce) yield class 18 m³/ha may be determined.

Discounted Income

Using the price/size relationships of Table 26, discounted revenue can be calculated at various interest rates for the different species and yield classes. The calculations are very similar to the ones already demonstrated for discounted expenditure. A detailed analysis will be shown for Sitka spruce yield 18 m³/ha assuming constant prices.

TABLE 26. *Picea sitchensis* Y.C. 18 m³/ha Standing Volume and Value of the Crop

Age (years)	Main crop		Thinning	
	Volume (m³/stem)	Price (£/m³)	Volume (m³/stem)	Price (£/m³)
17	0.06	3.0	0.03	2.0
20	0.11	4.0	0.05	2.5
23	0.18	5.0	0.10	3.5
27	0.33	6.5	0.18	4.5
31	0.52	8.5	0.28	6.0
35	0.78	10.5	0.38	8.0
40	1.12	12.5	0.63	10.0
45	1.51	14.5	0.83	12.0
50	1.94	16.5	1.00	12.0
55	2.50	16.5	1.12	12.0

With regard to volume, this is shown in Table 27, together with the total net price obtained for the timber. No allowance at this stage has been made for loss of volume due to damage by the elements and pests, neither has any allowance been made for roads and rides. The forest is considered to be 100% stocked. (The British Forestry Commission generally make an allowance of 15%: 10% for damage and 5% for rides.) The woodland manager must make his own assessment of the unproductive area per compartment and take account of this in the final financial yield calculation or reduce the volume production by a certain percentage, depending on the calculated loss. The former method is favoured in this chapter and is demonstrated later on. Again, it is unlikely that the volume production from a 100% stocked area will be

TABLE 27. *Volume and Money Yield: Sitka spruce Y.C. 18 m³/ha Price Gradient as per Table 25—no Inflation*

		Thinning per ha			Main crop per ha			Felling	
Age (yrs)	Stems (no.)	Volume (m³)	Value (£)	Value as felling (£)	Stems (no.)	Volume (m³)	Value (£)	Volume (m³)	Value (£)
17	(630)	20	40	(50)	1750	100	300	120	350
20	390	20	50	(60)	1360	145	580	165	640
23	300	30	105	(120)	1060	195	975	225	1095
27	280	50	225	(250)	780	255	1658	305	1908
31	180	50	300	(325)	600	310	2635	360	2960
35	130	50	400	(425)	470	365	3832	415	4257
40	95	60	600	(630)	375	420	5250	480	5880
45	60	50	600	(625)	315	475	6888	525	7513
50	45	45	540	(562)	270	525	8662	570	9224
55	40	45	540	(562)	230	575	9488	620	10,050

Note:
(i) Planting distance 2 m; 2500 p.p.ha.
(ii) The felling totals in Table 27 are the volume and value of the crop assuming it is clear felled at the specific age. They are the summation of the thinning and main-crop volume and value for the year in question, care being taken to increase the "thinning" value to that of the felled crop to take account of the price difference between the two (see Table 25).

exactly as shown in management tables. The manager will again have to make the necessary adjustments.

It must be remembered that when calculating the money yield from a plantation, the income from *each thinning must be discounted separately before adding to the discounted felling value* because these incomes are obtained at different time periods.

The formula used for discounting revenue is exactly the same as that used for discounting expenditure, namely

$$Vo = Vn \left(\frac{1}{(1.0p)^n}\right) \text{ and the multiplier } \left(\frac{1}{(1.0p)^n}\right)$$

for specific years and specific rates of interest can be obtained from compound interest tables. Table 28 lists the multipliers used.

The multipliers given in Table 28 for the specific rates of interest and the specific age may be used to give the discounted values of the incomes shown in Table 27. For example, at 5% and year 20 the discounted thinning value is equal to £19 (£50 × 0.377). Likewise the discounted felling value equals £241 (£640 × 0.377). In a similar way all the

TABLE 28. *Multipliers used to calculate Discounted Income*

Age (yrs)	Rate of interest (%)			
	5	7	10	12
17	0.436	0.316	0.198	0.146
20	0.377	0.258	0.147	0.104
23	0.326	0.211	0.112	0.074
27	0.268	0.161	0.076	0.047
31	0.220	0.123	0.052	0.030
35	0.181	0.094	0.036	0.019
40	0.142	0.067	0.022	0.011
45	0.111	0.048	0.014	0.006
50	0.087	0.034	0.009	0.003
55	0.068	0.024	0.005	0.002

discounted thinning and felling values may be computed and this has been done in Table 29.

It would be as well to study Table 29 in some detail. The discounted income for the thinnings "running total" column is simply the summation of the thinnings discounted income column. For example, at 5% and 20 years £17 + £19 = £36, and so on. Although the real value of the thinnings and final felling increase all the time (Table 27), the discounted incomes at first increase, reach a peak and then start to decrease (Fig. 4). The peak is attained at an earlier date, the higher the rate of interest. For example, in the D.I. final felling section, the peak is 40 years at 5%; 7%—35 years; 10%—31 years and 12%—27 years.

The total discounted income columns are obtained by adding to the final felling discounted income the discounted income from previous thinnings. Thus, at 5% interest rate and a rotation of 23 years the total discounted income is £357 + £36 = £393.

The maximum rotation shown in Table 29 is 55 years. If the trees were allowed to grow for much longer the final felling would contribute practically nothing to the discounted income total. Even at 55 years £10,050 only contributes £20 to D.I. total at 12%, the remaining £59 of £79 being derived from thinnings revenue. Irrespective of final income, after a certain age the discounted value of this income will in effect be zero. Figure 4 illustrates the point.

For a rotation of greater than 70 years at 10% rate all the discounted income is made up from thinnings revenue. It will be seen from Fig. 4

TABLE 29 *Discounted Income from Thinnings and Final Felling. Sitka spruce, Y.C. 18 m³/ha. Single-rotation constant prices*

Thinning		D.I. thinning value Rate of interest (%)				D.I. thin running total value					Felling		D.I. final felling value Rate of interest (%)				D.I. fell. and thin value Rate of interest (%)			
Age (yr)	Value (£)	5	7 (£)	10	12	Years	5	7 (£)	10	12	Age (yr)	Value (£)	5	7 (£)	10	12	5	7 (£)	10	12
17	40	17	13	8	6	17	17	13	8	6	17	350	153	111	69	51	153	111	69	51
20	50	19	13	7	5	17–20	36	26	15	11	20	640	241	165	94	67	258	178	102	73
23	105	34	22	12	8	17–23	70	48	27	19	23	1095	357	231	123	81	391	257	138	92
27	225	60	36	17	11	17–27	130	84	44	30	27	1908	511	307	145	90	581	365	172	109
31	300	66	37	16	9	17–31	196	121	60	39	31	2960	651	364	155	89	781	448	199	119
35	400	72	38	14	8	17–35	268	159	74	47	35	4257	771	400	153	81	967	521	213	120
40	600	85	40	13	7	17–40	353	199	87	54	40	5880	835	394	129	65	1103	553	203	112
45	600	67	29	8	4	17–45	420	228	96	58	45	7513	834	361	105	45	1187	560	192	99
50	540	38	13	3	1	17–50	458	241	99	59	50	9224	802	314	83	28	1222	542	179	86
											55	10050	683	241	50	20	1141	482	149	79

Note: The underlined values are the peak values at specific rates of interest.

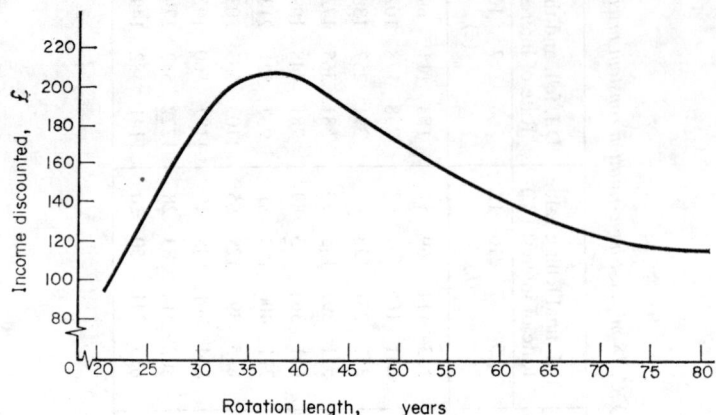

Fig. 4. Discounted income S.S. Y.C. 18 m³/ha, constant prices, 10% discount rate.

that if the discounted expenditure came to £193 the woodland manager could earn at least 10% on invested money if he felled the crop between 32 and 44 years of age. However, as will be shown in Chapter 14 it is not necessarily true that he will maximise his return on invested money by felling the crop at year 35—the peak of the 10% discounted income curve. It may be that the 11% curve peaks at year 30 and this maximum value is just sufficient to cover discounted expenditure; this then will be the optimum rotation and the maximum return.

Chapter 13

FINANCIAL YIELD[1]

Introduction

How are the optimum rotation and the maximum financial return determined? A series of curves can be drawn showing discounted income for specific rotations at different rates of interest. This is shown in Fig. 5 for Sitka spruce Y.C. 18 m³/ha. The curves represent this discounted income for one rotation only (Table 29). Also plotted on the graph is the discounted expenditure curve (Table 21). It can be seen that whereas the discounted-expenditure curve is gently sloping, the discounted-income curve slopes very steeply to start with and then gradually tails out. A close inspection of the graph shows that after 9.5% interest rate the 40-, 45-, 50- and 55-year "rotation" curves cross and fall below the 35-year rotation curve. This cross-over means that peak discounted income is reached at an earlier and earlier age as the rate of interest increases. In other words, unless there is a very abnormal price/size gradient, the higher the rate of interest the woodland manager wishes the plantation to earn, the shorter the rotation must be and the earlier he has to fell, the *optimum rotation being where the discounted expenditure cuts the outer discounted income line.* In Fig. 5 the optimum rotation length is about 35 years and the financial return for this rotation is approximately 10.4%.[2]

[1] Also known as Internal Rate of Return (I.R.R.).

[2] Another method of determining Financial Yield (F.Y.), without using a graph, is by interpolation, Formula 9.

Formula 9: Financial yield determination by interpolation.

One discount rate is found—the lower discount rate (L.D.R.)—which gives a positive present worth (P.W.) value = discounted revenue minus discounted costs (∴ D.R. > D.C.). Another rate is then found—the higher discount rate (H.D.R.) —which gives a negative present worth value (D.R. < D.C.). This brackets the true F.Y. which may then be estimated by straight-line interpolation. The interpolation rule is:

$$\text{F.Y.} = \text{L.D.R.} + \left[\text{H.D.R.} - \text{L.D.R.} \left(\frac{\text{P.W. @ L.D.R.}}{\text{P.W. @ L.D.R.} - \text{P.W. @ H.D.R.}}\right)\right].$$

Example 14 illustrates the use of this formula (see footnote continuation, p. 144).

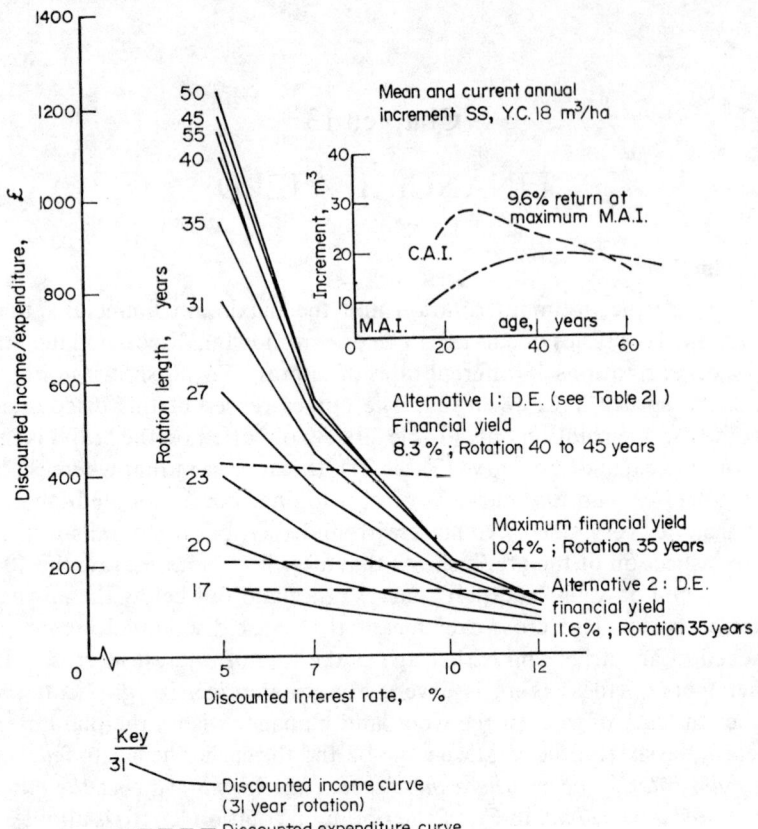

Fig. 5. Discounted-income and expenditure curves—one rotation only. Sitka spruce yield 18 m³/ha (constant costs and prices).

Example 14: *Determination of financial yield percentage*
Present worth @ 10% discount rate on a 35-yr rotation = £20 (213–193).
Present worth @ 12% discount rate on a 35-yr rotation = minus £70 (120–190).
$$\text{F.Y.} = 10\% + \left[12\% - 10\% \left(\frac{20}{20+70}\right)\right] = 10\% + \left[2\% \left(\frac{20}{90}\right)\right] = 10 + 0.44\% = \underline{10.44\%}.$$

The greater the difference between the two discount rates the more approximate the financial-yield determination becomes. Ideally the difference should not be more than 1 or 2%. Even so the true financial yield will be slightly less than that determined by straight-line interpolation for the points of the present-worth line usually lie on a concave curve.

If the expenditure had been double that, i.e. £465 (alternative 1, Table 21, p. 132) then the yield would have been 8.3% on a 40- to 45-year rotation. Similarly a cost of £179 (alternative 2) gives a financial yield of 11.6% on a 35-year rotation. It may seem strange *but the cheaper it is to plant a unit area, the earlier one gets the final return*, and, of course, the higher the financial yield becomes.

As a matter of interest, the mean and current annual increments for S.S. Y.C. 18 m³/ha have been plotted on the same graph as the above curves (Fig. 5). It will be noted that the C.A.I. curve peaks at age 25 years, but the maximum M.A.I. per hectare is not reached until about 55 years; this is after the maximum financial rotation. It need not necessarily be so, but it is the usual case. (It all depends on the price/size gradient.) If the policy of the enterprise is to fix the rotation at maximum M.A.I., then there usually will be a sacrifice of money return. Likewise at the point of maximum financial yield there will probably be a volume sacrifice.

Table 29 and Fig. 5 give the discounted revenue curves for one rotation only. As explained previously, it is incorrect to compare different "single" rotation lengths. This is overcome by discounting each rotation back from infinity by assuming that the gross revenue will be the same for the first and subsequent rotations. Exactly the same formula, Formula 8, is used as in the case of discounted expenditure, namely,

$$Vo = Vn \left(1 + \frac{1}{(1.0)^n p - 1}\right)$$

where Vo = discounted income from an infinite number of rotations of length n;

Vn = discounted income for one rotation;

p = rate of interest in per cent, and, therefore, the same multipliers (Table 23) are used.

Using the discounted income totals of Table 29 and the multipliers of Table 23, discounted incomes for an infinite number of rotations can be calculated. For example, at 5% interest and on 35-year rotations, the discounted income becomes £1180 (£967 × 1.22). Table 30 and Fig. 6 give the calculated discounted income curves for Sitka spruce Yield Class 18 m³/ha, assuming an infinite number of rotations.

It will be noticed, if Table 29 and Table 30 are compared, that in the 10% and 12% columns there is very little difference between the

TABLE 30. *Discounted Income for an Infinite Number of Rotations Sitka Spruce Y.C. 18 m³/ha (constant prices)*

Age (yrs)	Rates of interest (%)			
	5	7	10	12
17	271	162	86	60
20	413	240	119	80
23	582	326	156	99
27	796	422	186	114
31	1000	510	211	<u>123</u>
35	1180	573	<u>221</u>	122
40	1291	<u>592</u>	207	113
45	<u>1341</u>	588	194	100
50	1332	563	181	86
55	1221	492	150	79

two discounted values. One interesting point is that the curves for each "Rates of interest" column tend to peak at an earlier age when discounted back from infinity—Table 31.

TABLE 31. *Comparison of Maximum Discounted Income between S.S. Y.C. 18 m³/ha for a Single Rotation and in Perpetuity (constant prices)*

		D.I. total (maximum) Rates of interest (%)			
		5	7	10	12
Rotation age (yr)	Single rotation	50	45	35	35
Maximum D.I. (£)		1222	560	213	120
Rotation age (yr)	Perpetuity	45	40	35	31
Maximum D.I. (£)		1341	592	221	123

The discounted-expenditure curves are plotted on Fig. 6 as well as the discounted-income lines. Unlike Fig. 5, when one D.E. curve represented all the rotation ages, each rotation has a separate curve and three have been traced for D.E., the values being obtained from Table 24, p. 136. Where one specific discounted-revenue curve cuts the same specific discounted-expenditure curve, the financial return at that rotation can be read off the graph. Because of the nature of both sets of curves it does not necessarily follow that the outer discounted-income curve will give the maximum financial yield, although it is usually the case. Figure 7 illustrates the above point.

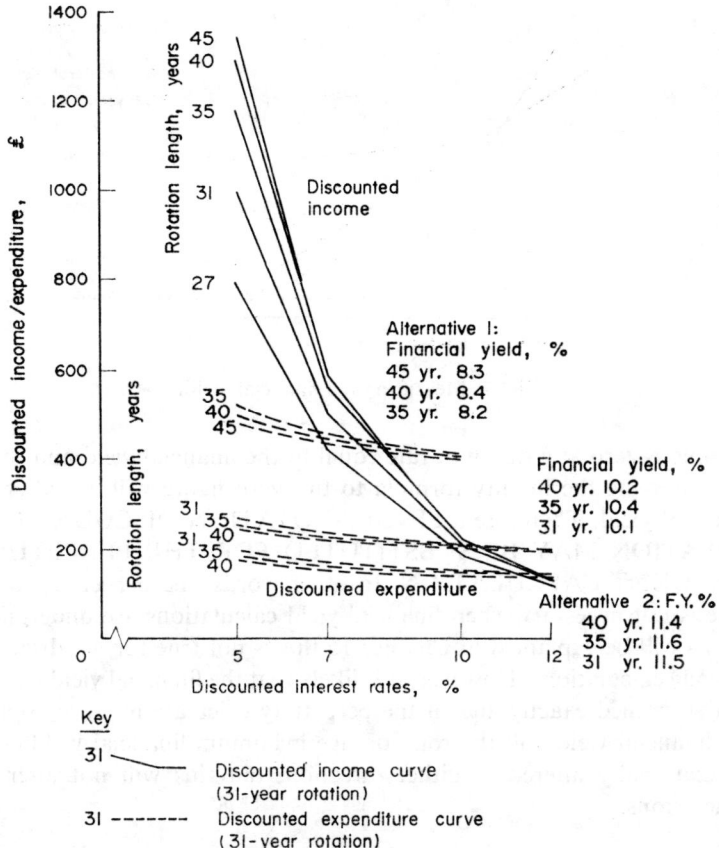

Fig. 6. Discounted-income and expenditure curves (discounted back from infinity). Sitka spruce Y.C. 18 m³/ha (constant costs and prices).

At a rate of interest of $Y\%$ the maximum discounted income will be given by the 35-year "rotation" curves, but as can be seen from the figure this curve does not give the maximum financial yield.

The financial yields for specific rotations at the different discounts rates are given on the diagram. When this diagram (Fig. 6) is compared with Fig. 5 it is seen that the rates of interest earned on the investments (the financial yield) are similar. This is logical for the net discount

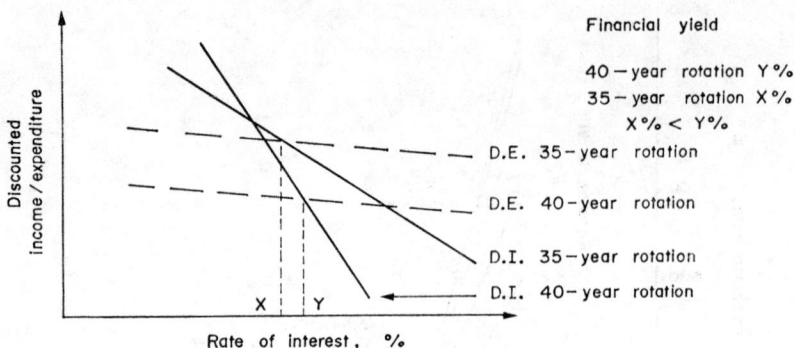

Fig. 7. Measuring the financial yield.

revenue is zero at a discount rate equal to the financial yield and therefore applying the infinity formula to this zero figure will not alter the financial yield. Therefore a SINGLE-ROTATION DISCOUNT CALCULATION MAY BE SUBSTITUTED FOR THE PERPETUITY DISCOUNT CALCULATION. In other words, the perpetuity allowance is not necessary when financial-yield calculations are undertaken, but as will be explained in Chapter 15 this is not true for net discount-revenue calculations. However, it is likely that the financial yield will not be determined exactly and, if the perpetuity calculations were applied, the financial yield and the rotation age maximum financial yield could be marginally altered in either direction, but this will not alter the conclusions.

Variations of Costs and Receipts on Financial Yield

Examples have been given of the financial yield from three sets of costs. From these it is seen that the higher the costs, all other things being equal, the lower the yield and the longer the rotation to give the maximum financial return. On the income side, if there is not a marked price gradient for the various size classes of trees, then the "younger" wood becomes relatively more valuable than in the example already described. The discounted-income curves are flatter and the financial yield will peak at an earlier age. Various cost and price structures can be tried by the manager to demonstrate the relative importance of each cost component and to give an idea of the market to aim for, whether it be fuelwood, pulpwood, sawnwood, etc., or a combination.

Inflation and Financial Yield

What effect does inflation have upon the expected financial yield? In most enterprises general inflation would have a negligible effect because expenditure and income from a particular enterprise would be affected equally. This is not the case in timber production or other long-term investments. The bulk of the capital expenditure in forestry takes place at the formation stage of a plantation whereas income does not usually start to flow until after 10 years in the Tropics or about 20 years in a temperate climate but the bulk of the income comes at the end of the rotation. Therefore, if the general price level of roundwood increases at the same rate as inflation, the financial yield will also increase by *at least the same average amount*; this is irrespective of whether the inflation rate started at a high level and decreased over time or vice versa. This phenomenon is best illustrated with a simple example. Table 32a gives the expected costs and revenues of a plantation on a 23-year rotation assuming current price levels.

If inflation took place over the lifetime of the project at an average

TABLE 32a. *Expected Costs and Revenues of a Plantation on a 23-year Rotation (S.S. yield class 18 m^3/ha)*

Units: £ per ha

Year	Cost (−) Revenue (+)	Discounted costs and revenues	
		8%	9%
0	−182[a]	−182	−182
2	−6	−5	−5
9	−4	−2	−2
16	−29	−8	−7
17	40	11	9
20	50	11	9
23	1095	186	151
Total	964	11	−27

F.Y. = 8.3%

[a]An allowance has been made for all overhead costs and included in the first year's cost.

rate of about 9%, then what would be the financial yield? Three models are built up with differing inflation rates, but each with the same average. The assumed inflation rates are shown in Table 32b.

TABLE 32b. *Inflation Rate per Year in per cent*

Year	Model A	B[a]	C[a]
0	0	0	0
1–5	9	16	4
6–10	9	10	6
11–15	9	8	10
16–20	9	5	13
21–22	9	4	13
23	9	5	17

[a] In models B and C the arithmetic average is slightly more than 9%; this is to ensure that the compound multipliers for each model are more or less equal in year 23.

Table 32c gives the costs and revenues for the different models and works out the financial yield.

It shows that the return of the forest project has increased by a rate greater than the average inflation rate; this is because investments in year zero are not inflated. Therefore, if timber prices keep pace with inflation the actual return to forestry will be the calculated *real* return plus the average inflation rate. This is a principle that has not been fully understood by forest officers when they have been putting up a project for consideration by governments. If current costs and prices are used to work out the financial yield of a plantation or to undertake project appraisal then the present inflation rate should be added to the financial yield if it is being compared with the actual returns from industrial and other projects. If this is not done, like is not being compared with like; this is the main reason why investment in forestry has appeared to give a low return.

The above argument was based on the assumption that timber prices will keep pace with inflation. Historically, timber prices have tended to rise faster (about $1\frac{1}{2}$% more p.y.) than other commodities; that is increase relatively faster than the average inflation rate. These measurements have been made over periods of 50 years or more.[3] Between 1972

[3] For a documented argument see Bradley, Grayson and Johnston, *Forest Planning*, p. 510, Faber & Faber.

TABLE 32c. *Financial Yield after Inflation S.S. Yield Class 18 m³/ha*
Units: £ per ha

Year	Original cost (−) revenue (+)	Inflated cost (−) and revenue (+) Model			Discounted costs and revenues								
					A		B			C		
		A	B	C	17%	18%	17%	18%		17%	18%	
0	−182	−182	−182	−182	−182	−182	−182	−182		−182	−182	
2	−6	−7	−8	−6	−5	−5	−6	−6		−4	−4	
9	−4	−9	−12	−6	−2	−2	−3	−3		−1	−1	
16	−29	−115	−151	−85	−9	−8	−12	−11		−7	−6	
17	40	173	218	134	14	10	18	13		11	8	
20	50	280	317	242	14	10	16	12		12	9	
23	1095	7947	7961	7906	246	175	247	175		245	174	
Total	964	8087	8143	8003	76	−5	78	−2		74	−2	
Financial yield after inflation					17.94		17.98			17.97		
Financial yield before inflation					8.29		8.29			8.29		
Difference due to 9% inflation					9.65		9.69			9.68		

and 1974 world timber prices doubled, whereas the average inflation rate was about 15% per year. This phenomenon could be partially explained by the fact that timber—a renewable resource—is competing with a group of non-renewable resources which are getting relatively scarcer. Secondly, the demand for timber products increases with wealth. The income elasticity of demand for most timber products is greater than one, therefore a 1% increase in wealth will cause a proportionately larger demand (1% plus) for the product. However, in 1975 timber prices fell to a level about 65% of the 1974 prices and they have remained more or less constant in *real* terms since then. This coincides with a general world recession, but the trend in timber prices semes to be on the increase again. If timber prices continue to rise faster than the general rate of inflation, then the real return will be more than the calculated return by an amount equal to the relative increase in price. One thing that is fairly certain is that timber prices will at least keep pace with inflation and therefore are a very good hedge against it.

Productivity may increase at a faster pace than labour rates and other costs. Therefore, as is to be expected, the real return will be greater than the calculated return which assumed constant costs and prices. On the other hand, if costs rise faster than productivity increase the reverse will occur.

Inflation and Its Effect on Rotation Length

The effect of inflation on rotation length presents rather a hazy picture. It all depends on the price/cost relationship. If the anticipated financial yield is relatively small, inflation may lengthen the rotation of maximum yield. Generally speaking, however, inflation tends to shorten marginally the rotation of maximum financial return.

Marginal Increase in Financial Yield

The manager may not think it worth while to alter the silvicultural practice, or try new species if only a marginal increase in financial return is expected. However, even a marginal increase over a long time span can amount to a substantial sum of money. If, for example, the manager could increase his anticipated return, say, from 10% to 10.1% on a 35-

year rotation, this would give an increased income of 0.9 unit for every unit invested. (£1.00 @ 10% for 35 yr = £28.10; £1.00 @ 10.1% for 35 yr = £29.01) or over 3% increase on actual money yield.

Drawbacks to selecting Rotation of Maximum Financial Yield

A marginal increase in financial yield may not always bring in the greatest quantity of money. Indeed, as stated previously, if maximum financial yield occurs before maximum mean annual increment, the manager is foregoing volume increments for financial gain. A 10.1% return on a 30-year rotation gives £18 on every £1.00 invested whereas a 10% return on 35-year rotation gives £28 per £1.00 invested. Clearly, if the whole £18 cannot be reinvested at 10.1%, the "second best" solution may be more profitable (seven 30-yr rotations @ £18 = £126, whereas six 35-yr rotations @ £28 = £168).

The manager may only be able to replant the area felled and have no alternative investment. He therefore should compare the money yield of the various rotation ages. For example, a yield of £28 for every £1.00 invested on a 35-year rotation is equivalent to a money yield of £24 on a 30-year rotation and £32 on a 40-year rotation.[4] The respective financial returns are 11.2% (30 yr), 10.0% (35 yr) and 9.0% (40 yr). If the manager cannot achieve the alternative rates at the specified rotation ages then, in spite of higher "financial yield" per unit area, he should aim for the highest money yield, bearing in mind the different rotations.

The manager has to consider all the factors such as demand for ready money, expectation rate, future market trends, government policy, etc., before deciding the immediate and long-term objects of management.

[4] To calculate these totals it has been assumed that at the end of each rotation only one unit (£1) could be reinvested in forestry on the existing land, no other land being available. Therefore to compare three different rotation periods, namely 30, 35 and 40 years, on an equal basis one has to compare 28 rotations of 30 years with 24 rotations of 35 years and 21 rotations of 40 years. If one 35-year rotation yields £28 for every £1 invested, 24 rotations will yield £672 per £. Similarly, 28 rotations of 30 years will require £24 to yield £672 and 21 rotations of 40 years will require £32 to yield £672.

Chapter 14

FINANCIAL YIELD OF THE ENTERPRISE

Introduction

The financial yield for a unit area, single species and yield class has been demonstrated. This should now be repeated for the various species and yield classes. Once this has been completed, all that remains to be done is to compile an estimate for the whole enterprise. This can be done quite easily if the area planted by species is known, and an estimate has been made of the expected yield class of every area.

Allowance for Crop Damage, Roads and Rides

In the example no allowance was made for understock areas, roads, crop damage, etc. This would be unnecessary if the crops had been actually measured, but if the assessment had been made from yield tables allowances may have to be made. Suppose that this has been estimated at 15%. Allowance has to be made for this loss, but it cannot be done by simply taking 15% of the financial yield (say 10.4%). This, because of the nature of compound interest, would give an over-estimate of the "loss" and, therefore, an under-estimate of the yield. It is necessary to calculate the return an investment would give at $X\%$ (say 10.4%) for Y years (35 years), and then subtract the loss, $Z\%$ (15%), from this total and find out what this new value will yield.[1] This is a straight calculation involving the compound-interest formula $V_n = V_o (1.0p)^n$.

[1]This is not strictly correct because it is assumed that there are no immediate money returns, just a final yield. However, it is a good approximation.

FINANCIAL YIELD OF THE ENTERPRISE

For Sitka spruce Y.C. 18 m³/ha,

$V_o = £1 \quad V_n = ? \quad p = 10.4\% \quad n = 35,$
$V_n = \log 1 + 35 \times \log 1.104,$
$V_n = £31.91,$
V_n less $15\% = £27.12.$

Apply the compound-interest formula again to find the new value of $p\%$:

$$V_n = V_o(1.0p)^n \quad \left(\frac{V_n}{V_o} = (1.0p)^n\right),$$

$£27.12 = 1(1.0p)^{35},$
$\log 27.12 \div 35 = \log 1.0p,$
$1.0p = 1.099,$
$p = 9.9\%.$

(Note that a straight 15% of $10.4\% = 8.8\%$.)

Determination of Expected Financial Yield for the Whole Enterprise

The expected financial yield for the forestry estate can be worked out by multiplying each expected yield by the appropriate area, adding the totals and dividing by the planted areas, as shown in Table 33a.

TABLE 33a. *Financial Return, Arbor Enterprise as at 30th September 1978*

Area (ha)	Species	%	Ha × %
1030	S.S.	9.9	10,197
180	N.S.	9.0	1620
150	S.P. & J.L.	3.7	555
1360			12,372
	$12{,}372 \div 1360 = 9.1\%$		

The expected return on gross capital expenditure for Arbor Enterprise (forestry section) is estimated to be 9.1% (constant costs and prices, no inflation, one rotation and 15% loss to the crop. No allowance need be made for the perpetuity factor but inflation, both general and relative (say 9.0%), has to be considered. Therefore, the expected financial yield is increased to 18.1% and the rotation ages of the crop may be reduced by 1 or 2 years.

In the example (Table 33a), the financial return could be increased if the Scots pine and the Norway spruce crops were replaced by Sitka spruce. It may pay the manager to foreshorten the rotation of the Scots pine and sacrifice some loss of revenue in order to obtain a great future return. This could be done by adopting a heavy thinning technique or by just felling at an earlier age. Using the method of discounted interest, the manager should calculate if such a policy would pay.

Table 33a gives the financial return for Arbor Enterprises for one set of assumptions. Similar calculations can be made by substituting the other variables, such as different cost and price assumptions. By doing this the manager will come to understand the importance of the different components and learn in what area the best efforts for increasing profitability can be applied.

These financial yields can be compared with the returns from industry in general, and from specific enterprises in particular. It is important to note that if the discounting process is being used to compare the returns from two mutually exclusive ventures—a forest plantation or a sheep farm—or where financial priority has to be established—extending the forest area or establishing a commercial nursery or building a sawmill—the time period chosen must be the same for each alternative. The manager must be careful when ascertaining the returns from forestry with those from other industries to compare the "after-tax" dividend because the taxation calculations may differ from industry to industry.

Calculation of Actual Financial Yield

The method of calculating expected financial yield has been demonstrated. This gives the woodland manager an idea of the profitability he can expect from certain species and yield classes and it tells him the optimum rotation. By making various assumptions about crop treatment, he can determine whether or not it will pay him to carry out this certain treatment. He can also determine the minimum price obtainable in order to give a certain return on capital. However, none of this necessary information will be available unless a detailed record is kept of costs and revenues.

Such a record is also important for calculating the actual financial yield of a plantation over its lifetime. An area (2 hectares) of Compart-

FINANCIAL YIELD OF THE ENTERPRISE

ment 11 in Arbor Enterprise was felled in 1978. Suppose that detailed records had been kept of the costs and receipts of Compartment 11, as per Compartment Operation Card (see Tables 7 and 11). Table 33b gives a summary from such a card for the compartment in question.

TABLE 33b. *Extract from the Compartment Record Card*

Compartment 11 1925 year planted			Area planted: 2 hectares Total area: 2 hectares
PER HECTARE COST AND REVENUE INCLUDE ASCRIBABLE OVERHEADS			
Year	Age		Cost
1925	0	Initial Establishment (S.P. planted at 1.2 m × 1.2 m Direct Notch. Planted on area felled in 1921)	£37.00
1927	2	Beating Up and Weeding	3.00
1929	4	Weeding	1.25
1937	12	Early Cleaning	0.62
1950	25	Brashing	10.00
1925–1950	26 Yrs.	Per hectare Overheads per year	0.62
			Revenue
1951	26	1st Thinning	10.00
1956	31	2nd Thinning	30.00
1961	36	3rd Thinning	35.00
1965	40	Windblow	3.00
1966	41	4th Thinning	40.00
1970	45	5th Thinning	45.00
1974	49	6th Thinning	50.00
1978	53	Clear felling	2400.00

Note: These figures do not take into consideration grants and tax rebates. However, inflation is built in, of course.

The compound-interest formula is used to calculate the financial yield. Seeing that the final felling took place in 1978, the yield can be determined either by compounding up to the present day (1978) or discounting back to the year of planting. When using the compound-interest formula $V_n = V_o(10p)^n$, it must be remembered that n refers to the year 1978. The multiplier $(1.0p)^n$ can be obtained from compound-interest tables, care being taken when calculating the annual overhead costs to obtain the correct multiplier. If the formula

$$V_n = R\left(\frac{1.0p^n - 1}{0.0p}\right)$$

is used, where R = annual overhead cost, the value V_n required = $V_{\text{year 53}} - V_{\text{year 27}}$. If the costs and revenues in Table 33b are compounded up, the actual financial yield amounts to 8.05%.

Discussion

Part III has demonstrated the use of the discounting (and compounding) process in order to work out the expected yield and optimum rotation age from an individual plantation and from the whole forest estate. The influence of various capital costs, different price/size gradients and inflation has been examined. It has been demonstrated that the lower the capital cost, the *earlier* and the higher the expected return should be. Again, if the price/size gradient depended on felling and extraction costs alone, the optimum rotation would peak at an earlier date than if the price of timber increases with diameter as well. Because the bulk of capital costs occur at the outset of a plantation life, whereas the revenue does not start to accrue until 10 to 20 years later, inflation (both relative and general) increases the financial yield by at least the amount of the inflation. Also, the optimum rotation age has a tendency to shorten. When comparing two different crop rotations similar time periods should be chosen in order to make the comparison valid. However, as has already been stated (p. 148), if financial-yield calculations are undertaken then crops of different rotation ages may be compared without discounting back from infinity because the financial yield will be the same irrespective of rotation numbers, if the assumptions about constant costs and constant revenues are valid. This is not true for net discount revenue calculations.

When calculating financial yield the interest rate is varied. Another method—*Net discount revenue*—uses one rate of interest only when comparing the profitability of different species. This is described in the next chapter.

Chapter 15

NET DISCOUNT REVENUE[1]

Introduction

In financial-yield calculations discounted values of costs and revenues are represented in a series of values which decrease in amount as the rate of interest is increased. A simplified method is to choose just one rate of interest and compare costs and revenues, say for different rotations, fixing the rotation of maximum money yield where the net discount revenue—that is discounted revenue minus discounted cost—is at a maximum. It can be seen that this method is just a slimmer version of the financial-yield method but it has one less variable. This restriction brings certain drawbacks and these are discussed below.

Methodology

The method of determining "Net Discount Revenue" (N.D.R.) is exactly the same as for calculating "Financial Yield". However, whereas with the Financial Yield method several rates of interest are selected in order to determine the yield, with Net Discount Revenue only one interest rate is chosen, discounted income and expenditure being determined at this specific rate. Also, as the name implies, the revenue is *net*, that is income minus expenditure. It follows, therefore, that this process has far less calculations and is quicker to cary out than financial-yield computations.

Selection of Interest Rate

The biggest question mark confronting this method is what rate of interest to select? Only by chance or an inspired guess will the manager

[1] Also known as net present worth (N.P.W.).

choose a rate of interest that is the same as the maximum financial yield for a particular plantation. If he picks this rate of interest, then the net discount revenue—the difference between the discounted income and expenditure—will peak at zero. If the interest rate is above the maximum then N.D.R. will be negative, and vice versa.

Internal Rate of Interest[2]

The manager, after consultation with the owner, may decide that profit maximisation is not the principal objective and that the enterprise will predetermine what it considers to be a satisfactory return on invested capital. This rate of interest, for convenience, shall be termed the internal (or owner's) rate of interest. It is defined as the rate of return that the enterprise will be satisfied with on its investments.

Let us examine what will happen if one specific interest rate is chosen. Table 34 gives the N.D.R. at various rates of interest and different

TABLE 34. *Net Discount Revenue per Hectare. S.S. Y.C. 18 m^3/ha (Single rotation, constant costs and prices)*

Unit = £

Rotation length (yr)	Rate of interest (%)			
	5	7	10	12
17	−53	−89	−124	−139
20	52	−22	−91	−117
23	187	57	−55	−98
27	375	165	−21	−81
31	575	248	6	−71
35	761	321	<u>20</u>	−70
40	897	353	10	−78
45	981	360	−1	−91
50	<u>1016</u>	<u>342</u>	−14	−104
55	935	282	−44	−111

Note: This table has been compiled by subtracting the discounted revenue (Table 29) from the discounted expenditure (Table 21).

---

[2] The internal rate of interest or owner's rate of interest should not be confused with the internal rate of return (I.R.R.) which is synonymous with financial yield (F.Y.).

rotation ages. Again Sitka spruce Yield Class 18 m³/ha has been used, and cost and prices as used in the financial-yield calculations assumed.

If the internal rate of interest is 7%, then the plantation could be felled at 23 years. Theoretically the return would be slightly in excess of 7% because the N.D.R. at age 23 is positive (£57.00). If inflation took place at 9% per year (general inflation 7.5%; relative inflation 1.5%), then the actual rate of return would be something like 16%. Similarly, if the internal rate of interest is 10%, disregarding inflation, etc., the rotation age would be 31 years.[3] For this particular specie and yield class, with the specific costs and prices assumed, N.D.R. is always negative at rates above 12%. Therefore, the "minimum" rotation age will vary directly with the selected rate of return on capital investment (internal rate). The smaller the rate of interest, the shorter the rotation will have to be in order to earn this "minimum" return on capital.

It is also possible that there will be a very large time period within which the plantation will earn at least the designated return on capital (internal rate of interest). For example, at 10% N.D.R. rate the rotation age ranges from 31 to 45 years. However, the rational forest manager will fell when N.D.R. is maximum for felling at that point will give the greatest capital gain.

Maximum Net Discount Revenue

A selected rate of interest may be chosen, but in place of determining the rotation age when the various crops have just earned the minimum return, the rotation age is fixed at the point when the net discount revenue is at a maximum. By adopting this method, i.e. maximising the net discount revenue (N.D.R.), a variation of the financial-yield calculation is being used, but it is only a "second best" solution. Unless a rate of interest is chosen that coincides with the financial yield, maximum discount will give a return less than the financial yield and a rotation greater than the financial optimum. For example, in Table 34 the maximum N.D.R.s have been underlined for the various rates of interest and these occur at 50, 45, 35 and 35 years, respectively, for 5, 7, 10 and 12% rates of interest. Therefore, if maximum N.D.R. is the policy, the

[3]Strictly speaking the yield at this stage is fractionally more than 10%.

lower the selected rate of interest, the longer the rotation and, of course, the higher the expected discounted revenue.

Using the above method when comparing profitability of two species, *conflicting results may occur*, depending on the rate of interest chosen. For example, consider an area that will grow two timber species equally well; both give the same total volume on the same rotation but one species gives a higher (and earlier) volume in thinnings, whereas the other has a greater final felling volume. If the interest rate chosen is a low one, and a comparison is made then the species with the high final felling volume is most likely to be favoured. However, a similar comparison but using a high interest rate may favour the species that gives a higher thinning volume. This apparent contradiction is perfectly logical because at higher rates of interest the "more valuable" early thinning returns weigh more heavily than the greater but later final felling returns. Figure 8 illustrates this point.

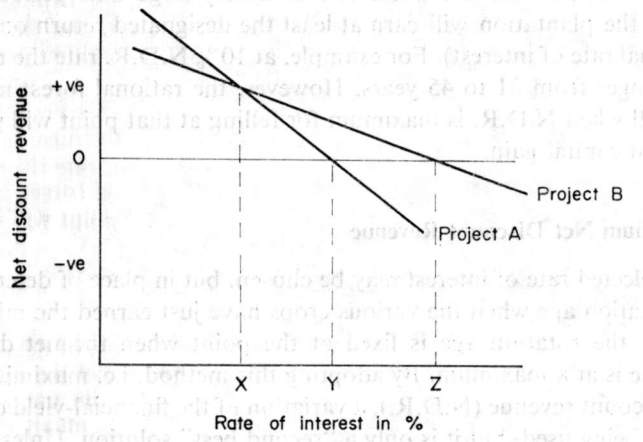

Fig. 8. Net discount revenue of two equal-aged projects at varying rates of interest.

At rates of interest below $X\%$ project A has a higher N.D.R. than project B, whereas at rates above $X\%$ B's N.D.R. is higher than A's. The financial yield of B is $Z\%$ which is greater than that of A ($Y\%$). Different projects will be favoured using the N.D.R. criterion depending on whether or not the selected discount rate is greater or less than $X\%$.

However, if the financial-yield criterion was chosen, only one solution is possible.

Again, whereas the financial yield is the same or only marginally different whether a single rotation or infinite rotations are used the net discount revenue can change significantly. This is shown in Fig. 9 for two projects of different durations. A reversal in the choice of projects may occur depending on whether the comparison is made between the projects on one rotation, Fig. 9a, or over an infinite number of rotations, Fig. 9b.

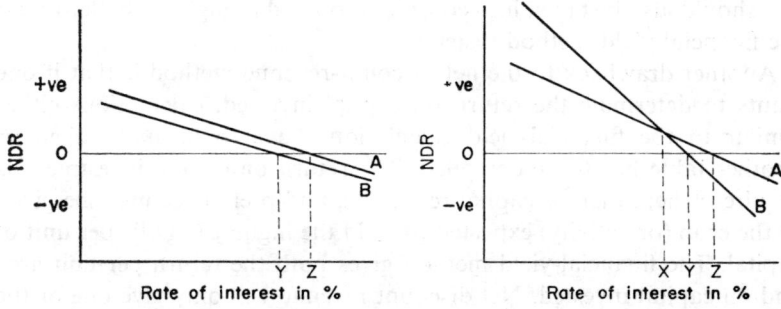

Fig. 9a Net discount revenue of two projects over a single rotation.

Fig. 9b. N.D.R. of two projects compared over an infinite number of rotations.

Project A lifetime n years
Project B lifetime m years
$n > m$

In Fig. 9a, project A has a higher N.D.R. at all rates of interest and therefore is the favoured project. In Fig. 9b when the two projects are compared over the same time period B has a higher N.D.R. at interest rates below $X\%$. However, the financial yield under either assumption is still the same. When the net discount-revenue criterion is used to compare projects with different maturity ages then they must be compared over the same time period to obtain meaningful results. As has already been stated in Chapter 13, p. 148, the single-rotation financial yield may be substituted for the perpetuity-discount calculation without affecting the conclusion. The cross-over and reversal phenomena are not uncommon and this is a serious weakness with the N.D.R. method.

Using the method of financial yield, it was shown that the cheaper the establishment and maintenance costs, the shorter the rotation for maximum financial yield and, of course, the higher the return. But, irrespective of capital costs, the net discount revenue will always peak at the same rotation age because it is concerned with total revenue over the period rather than the rate of return. Therefore, using this method full advantage will not be taken of cost reductions. That is it does not measure the efficiency with which the resource is being used. It is an absolute measure and it may produce a high N.D.R. just because the project is large. Therefore when comparing projects by this method the size should also be taken into consideration and it might be better to use the financial-yield method instead.

Another drawback to the net discount-revenue method is that if one wants to determine the return on capital invested, calculations either similar to the financial-yield calculations have to be undertaken or another table has to be compiled. The return on capital investment is required where there is capital restriction and preference must be given to the crop (or activity) expected to yield the highest N.D.R. per unit of capital. The financial-yield method gives both the return per unit area and on capital invested. Net discount revenue will only give one or the other.

Net discount revenue is not favoured in this manual as a method of determining profitability, for the reasons mentioned above, unless the owner or manager wishes to earn one specific rate of interest. Arguments in favour of its use and greater detail of how it is used may be found in three recent publications. These are:

1. *The Forestry Commission Working Plan Code*, H.M.S.O. (U.K.), 1970.
2. *Forest Planning*, R. T. Bradley, A. J. Grayson and D. R. Johnston, Faber & Faber, 1967.
3. *Practical Forestry for the Agent and Surveyor*, 2nd Edition, C. E. Hart, The Estates Gazette Ltd., 1967.

Chapter 16

INCREASING PROFIT

Market Intelligence

The forest manager should be constantly trying to find ways and means to increase the profitability of the enterprise. There are two broad ways in which he can set about this task: the reduction of costs and the application of market intelligence.

By market intelligence is meant having a knowledge of present national and local markets for the different wood-using industries. Such knowledge can be exploited in order to obtain the best going "net" price for the various categories of timber. Again, it should be used to forecast future trends and prospects so that a guide can be given to the manager in planning present plantings and replantings. This latter task is rather difficult because the market supply (and demand) conditions may have to be forecast 30 to 40 years ahead. However, broad trends can be determined and these may be employed by the manager to help him decide the species to plant.

The Food and Agricultural Organisation of the United Nations has issued a number of forecasts for various areas of the globe, as well as for the world itself.[1] These publications highlight a number of important trends in the consumption of timber products, and the manager may consult them for his particular country, as well as using import, export and consumption statistics of his own country. He could then use these figures for the different groups of forest products to make forecasts of future consumption, using such indicators as population growth and economic expansion.

---

[1] For example: *European Timber Trends and Prospects*, F.A.O., 1976; *Timber Trends and Prospects in Africa*, F.A.O., 1966; *Wood: World Trends and Prospects*, F.A.O., 1967.

Reduction of Costs

The manager can increase the income of the enterprise by correct market intelligence, but reduction of costs is perhaps the surest way to greater profitability. Of course, in order to reduce costs it is essential to have a knowledge of individual costs and of the specific components of each particular item. The manager can then judge where the greatest savings are to be made. If detailed costing information is not available, at best the manager will be making inspired guesses.

There are two principal ways of reducing unit costs, the first by increasing productivity on a particular operation—replacing hand draining by mechanical draining, for example—and secondly, by judicious and intelligent management—for example, decreasing the number of trees planted per unit area. In order to increase productivity it is important that the manager provides long enough runs of work of one kind so that efficient use be made of labour and machines; this may not be possible on small forest areas and, therefore, co-operation should be welcomed as a possible solution.

Fencing is the easiest operation to demonstrate the importance of economies of scale. Every time an area is quadrupled the cost per unit area will be reduced by half. Fencing is one of a number of establishment operations. Such operations occur at the start of a plantation's life and because of the relatively long waiting period between establishment and the first thinning income, there is a great need to reduce these costs to a minimum.

There are a number of ways of reducing establishment costs but the most obvious, and most economically desirable, is to increase the planting distance. Within certain limits it matters very little which planting distance is adopted; the end result will give the same volume per unit area but, of course, the wider the spacing the more volume there will be per tree (Fig. 10).

A spacing of 3 metres by 3 metres will result in a volume loss of between 10 and 20% of the maximum, and at 2.5 m by 2.5 m the loss will only be between 2 and 10%. The wider the spacing, the cheaper it becomes to plant an area, and when the plantation reaches the thinning stage, the cheaper it is to fell and extract per unit volume. The first point—cost of planting—can be illustrated as follows in Table 35.

INCREASING PROFIT 167

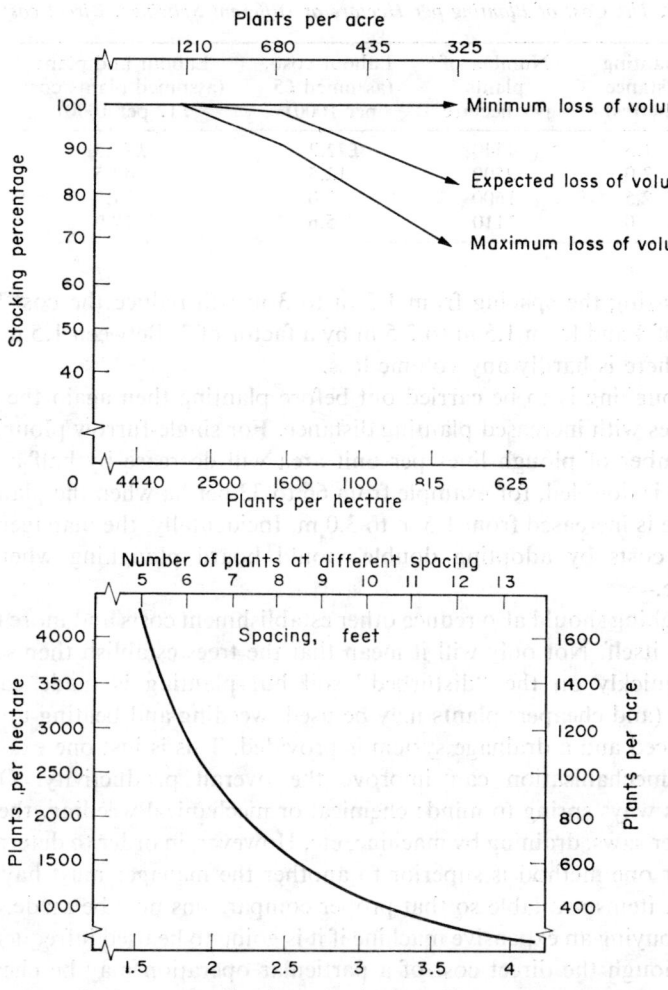

Source: Wardle, P. A., Spacing in plantation, *Forestry*, Vol. 40, 1967

Fig. 10. Relationship between spacing and volume production.

TABLE 35. *The Cost of Planting per Hectare at Different Spacings. Direct cost only*

Planting distance (metres)	Number of plants per hectare	Labour cost (assumed £5 per 1000)	Labour and plant (assumed plants cost £12 per 1000)
1.5	4440	£22.2	£75.5
2.0	2500	12.5	42.5
2.5	1600	8.0	27.2
3.0	1110	5.6	18.9

Increasing the spacing from 1.5 m to 3 m will reduce the cost by a factor of 4 and from 1.5 m to 2.5 m by a factor of 3. Between 1.5 m and 2.5 m there is hardly any volume loss.

If ploughing is to be carried out before planting then again the cost decreases with increased planting distance. For single-furrow ploughing the number of plough lines per unit area will decrease by half as the spacing is doubled, for example from 66 to 33 per ha when the planting distance is increased from 1.5 m to 3.0 m. Incidentally, the manager can reduce costs by adopting double mould board ploughing wherever possible.

Ploughing should also reduce other establishment costs and more than pay for itself. Not only will it mean that the trees establish themselves more quickly on the "disturbed" soil but planting is made easier, smaller (and cheaper) plants may be used, weeding and beating up will be reduced and a drainage system is provided. This is just one example where mechanisation can improve the overall productivity. Other obvious ways spring to mind: chemical or mechanical weeding, the use of power saws, draining by machine, etc. However, in order to determine whether one method is superior to another the manager must have all the cost items available so that proper comparisons may be made. It is no use buying an expensive machine if it is going to be used infrequently. Even though the direct cost of a particular operation may be cheaper with the machine, more likely than not the overheads will make this method the more expensive.

Wider spacing besides reducing establishment costs should lead to an increased income; trees will be of greater diameter/volume at a specific age, the wider the spacing. Even if there is no price gradient for increase in diameter, the larger trees are more valuable because they are cheaper

per unit volume to fell and extract. Suppose that in place of the 2-m spacing of the Sitka spruce Yield Class 18 m³/ha (Table 27), the trees had been planted at a distance of 2.5 m by 2.5 m, what would be the financial yield from the plantation? The number of trees per hectare is reduced by one-third; therefore, it is to be expected that each tree at the wider spacing will have, on average, 33% more volume. Even making a pessimistic assumption that there will be an 8% loss of volume and a loss in timber quality resulting in a much smaller price/size gradient increase than previously assumed, the financial return will be significantly greater.[2] This the manager can determine by undertaking for himself financial-yield calculations using his own costs and prices.

Likewise, he could assume that a more intensive thinning régime is to be adopted and calculate the financial yield. Within certain limits the more intensive the thinning the higher the financial yield.

Accurate costing is one of the principal tools of good management. It is carried out in order to determine profit or loss over the whole enterprise or in one particular sector. Once unit costs are known they can be used to budget, for fixing piece work and bonus rates and to compare other costs both within and outwith the enterprise. This book has tried to explain how the manager should attempt not only to cost the various forestry operations but how to use this information once it has been ascertained. Forestry has a place in most countries, but only through successful management will it be certain that its place is both secure and growing.

[2] An 8% loss in volume and a 7% loss in value of the price/size gradient will give a financial yield of 100% on a 35-year rotation as compared with 10.4% on a similar rotation at 2-m spacing (Fig. 6). With an optimistic assumption of no volume or quality loss the anticipated financial yield will be about 11.2%.

Appendix I
Standard Account Heads—Expenditure

Note

The time period for establishment and tending varies from country to country. This example illustrates conditions in the United Kingdom. Also the terms used are those generally applying in the U.K. This "English" terminology will have to be altered to suit the particular requirements of the individual countries.

0–9 Establishment (up to 5 yr in the U.K.)

Units to be used when costing

0 Fencing (a) Plain stock (sheep, goats, cattle, etc.)   Area and length
           (b) Rylock stock
           (c) Rabbit and other burrowing animals
    (a & c) Stock and rabbit
           (d) Roe deer (and other leaping animals)
           (e) Red deer

1 Draining (a) Main drains   Area and length—indicate if draining by (i) hand, (ii) machine
           (b) Side drains
           (c) Upkeep and maintenance as preparation for planting only

2 Ploughing (a) Single furrow   Area
            (b) Single furrow with tine
            (c) Single mould-board
            (d) Double mould-board
            (e) Discing

3 Clearing (a) Scrub   Area
           (b) Lop and top
           (c) Burning
           (d) Taugya/Shamba system

4 Manuring (a) Hand   Area and number (per 1000 trees)
            (b) Mechanical
            (c) Aerial

5 Other preparatory treatments
      (a) Turfing   Area and number (per turf)
      (b) Preparation for natural regeneration   Area
      (c) Elephant trenches, etc.

APPENDIX I

		Units to be used when costing
6	Artificial and natural regeneration (including screefing)	
	(a) Planting	Area and number (per 1000 trees)
	(b) Direct sowing	Area
	(c) Natural regeneration	
7	Weeding (up to 5 years)	
	(a) Hand	Area
	(b) Machine	
	(c) Spray	
8	Beating up (up to 5 years) (Includes replanting—100% B.U.)	Area and number (per 1000 trees)
9	Local or miscellaneous establishment operations	

Note

For establishment area referred to usually area planted.

10–19	*Tending (from 6th year onwards in the U.K.)*	
10	Early cleaning: until formation of canopy—removal of all adverse growth	Area
11	Late cleaning: after formation of canopy—removal of woody growth	Area
12	Beating-up (belated)	Area and number (per 1000 trees)
13	Enrichment planting (interplanting)	Area and number (per 1000 trees)
14	Underplanting	Area and number (per 1000 trees)
15	Manuring (belated)	Area
16	Rack-cutting (line cutting)	Distance
17	Brashing (a) 100%	Area
	(b) 50%	
	(c) 33%, etc.	
18	Pruning—high pruning by species	Area and number of trees
19		

20–39	*Harvesting and Conversion (in the Woods)*	
20	MARKING[1]	Number
21	MEASURING[1]	Volume
22	Thinning (a) Hand	Area, volume and stems
	(b) Machine	
23	Felling (of mature timber) (a) Hand	Area, volume and stems
	(b) Machine	
24	Underwood cutting (a) Hand	Area
	(b) Machine	

[1]Capital letters denote overhead costs.

			Units to be used when costing
25	Extraction to ride/road side and piling (if separate from 22–24)	(a) Hand (b) Animal (c) Mechanical (d) Machine	Area, volume and stems
26	Loading and transport	(a) Hand (b) Machine	Weight
27	Hauling to timber yard		Weight
28	Burning		Area
29			
30	Cordwood or long firewood		Volume or weight
31	Firewood		Volume and weight
32	Firewood blocks		Weight
33	Poles		Volume and number
34	Fencing material	(a) Posts (b) Strainers (c) Stakes	Volume and number
35	Decorative material	(a) Rustic fencing (b) Fencing rails (c) Hurdles	Volume and number
36	Commercial wood	(a) Pitprops (b) Pulpwood (c) Chipwood	Weight and volume
37	Christmas-trees		Each
38	Greenery and moss		Bundles
39			

Note: (i) after a number = snedding (branch removal), e.g. 22(i)
(ii) after a number = barking (peeling), e.g. 33(ii)
(iii) after a number = cross-cutting, e.g. 36b(iii)

If you cannot separate extraction cost, group two numbers together, e.g. 22/25.

40–49	*Repairs and Improvements—Woodlands (Overheads—Woodlands)*	
40	CONSTRUCTION OF WOODLAND BOUNDARY FENCES (where not for current planting)	Length
41	WOODLAND BOUNDARY HEDGES	Length
42	FENCING MAINTENANCE (except when preparation for planting—0)	Length
43	DRAINS UPKEEP (except when preparation for planting—1)	Length
44	ROADS CONSTRUCTION (depreciation 1/40th)	Length
45	ROADS UPKEEP (depreciation 1/40th)	Length
46	RIDES/TRACES (fire lines) CONSTRUCTION	Length
47	RIDES/TRACES (fire lines) UPKEEP	Length

APPENDIX I

		Units to be used when costing
48	FIRE DAM CONSTRUCTION AND UPKEEP (or POND)	Length
49		

50–59 *Protection (Overheads—Woodlands)*
- 50 FIRE-GUARD AND PATROL EXPENSES
- 51 FIRE-FIGHTING
- 52 PREPARATION OF FIRE BROOMS AND FIRE SIGNS
- 53 FIRE-FIGHTING EQUIPMENT (MAINTENANCE AND RENEWAL)
- 54 FIRE INSURANCE
- 55 VERMIN CATCHER'S WAGE (OR PROPORTION)
- 56 SUBSCRIPTION/PAYMENT TO VERMIN ERADICATION SOCIETY/ORGANISATION
- 57 CONTROL OF INSECT AND FUNGAL PESTS
- 58 STORES EQUIPMENT, ETC. (a) Cartridges
  - (b) Gun licence
  - (c) Gun repairs
  - (d) Traps
  - (e) Snares
  - (f) Poison gas
- 59

60–69 *Nursery*

60	Preparing new ground		Area
61	Seedbeds	(a) Preparation	Area
		(b) Sterilisation	
		(c) Sowing	
		(d) Weeding	
		(e) Watering	
		(f) Protection	
62	Seedlings	(a) Lifting	Number (1000 plants)
		(b) Grading	
		(c) Heeling-in	
		(d) Packing	
63	Lines/Pots	(a) Preparation/mixing soil	Area
		(b) Filling pots with soil	
		(c) Sterilisation	
		(d) Lining out/Planting in pots	
		(e) Weeding	
		(f) Watering	
		(g) Protection	
64	Transplants	(a) Lifting	Number (1000 plants)
		(b) Grading	
		(c) Heeling-in	
		(d) Packing	

173

174 COST AND FINANCIAL ACCOUNTING IN FORESTRY

			Units to be used when costing
	65	Delivery of plants	
		(a) Internal use	Number (1000 plants)
		(b) Outside	
	66	Composting (a) Sowing beds with green crop	Area
		(b) Adding artificial fertiliser	
	67	Seeds (a) Collecting (including cone collection)	Weight
		(b) Purchase	
		(c) Extraction	
	68	OVERHEADS NURSERY	
		(a) PATHS preparation and upkeep	Length
		(b) HEDGES preparation and upkeep	
		(c) FENCES (temporary—permanent) preparation and upkeep	
	69		
70–79		*Sawmill and Yard*	
	70	Cross cutting and/or sawing in yard	Volume
	71	(a) Sawing timber in mill	Volume
		(b) Stacking of material	
		(c) Seasoning	
		(d) Grading	
		(e) Measure	
		(f) Stocktaking	
	72	Preparing fencing material and miscellaneous materials	Volume
		(a) Posts	
		(b) Strainers	
		(c) Stakes	
		(d) Droppers	
		(e) Gates	
		(f) Water boxes	Number
	73	Preservation including preservative	Volume
	74	Firewood blocks, slabs and pulpwood	Weight
	75	Maintenance (a) Saws	
		(b) Tools	
		(c) Other machines	
	76	Motive power (a) Petrol/Diesel	
		(b) Electricity	
		(c) Oil/Grease	
	77	Transport from the yard	Distance
	78	Depreciation. List of vehicles and machines with their value at the start and end of the year	
	79		
80–89		*Vehicles, Machines (excluding sawmills), Horses and Hand Tools*	
	80	Wages of driver or horseman and mate (cross-entry 25, 27, 65, 77, 113)	

APPENDIX I

Units to be used when costing

- 81 Fuel, etc. (a) Petrol
  - (b) Diesel
  - (c) Oil
  - (d) Grease
- 82 Animal account—maintenance, upkeep, purchase, etc.
- 83 Maintenance (a) Repairs
  - (b) Tyres
  - (c) Spare parts
- 84 Liabilities (a) Road tax
  - (b) Road insurance
  - (c) Road driving licence
- 85 List of vehicles and machines with value at start and end of year and depreciation
- 86 Hire of vehicles and machines (cross-entry 1, 2, 43, 45, etc.)
  - (a) From outside
  - (b) From estate

(Road construction hire kept separate from other operations)

- 87 Carriage in or out
- 88 Tools (hand tools, knapsack sprays, etc.)
  - (a) New
  - (b) Repair
- 89

A—Woodlands   *Forest operation no.*

90–109 *Consumable stores, etc.*

- 90 Fencing material from outside   (0, 40, 42)
  - (a) Timber
  - (b) Wire
  - (c) Netting
  - (d) Nails, etc.
  - (e) Preservatives
- 91 Fencing material from sawmill (cross-entry)   (0, 40, 42)
- 92 Draining materials—culverts, clay pipes, etc.   (1, 43)
  - (a) except for road construction
  - (b) for road construction
- 93 Road materials (a) for construction   (44, 48)
  - (b) for maintenance
- 94 Plants from outside for planting and beating-up
- 95 Plants from nursery for planting and beating-up (cross-entry)

(5, 6, 8, 12, 13, 14)

- 96 Fertilisers for plantations   (4, 15)
- 97 Weedkillers   (3, 7, 10, 11, 45, 47)

		Forest operation no.
98	Vermin killers	(6, 8, 12, 12, 14, 57, 58)
	(a) Poison gas, powders	
	(b) Cartridges and gun licence	
	(c) Traps and snares	
	(d) Tree stump chemical	
99	Fire fighting equipment	

B—*Nursery*

100	Cones and Seeds (cross-entry)	(61, 67)
101	Bought in plants	
	(a) Seedlings	(63)
	(b) Transplants	(63)
102	Chemicals (a) Fertilisers (cross-entry)	(60, 61, 63)
	(b) Sterilisers	
	(c) Weedkillers	
	(d) Sprays	
103	Materials (a) Sand/soil	(60, 61, 63)
	(b) Fencing material	(68)
	(c) Road material	(68)
104		

C—*Sawmill*

105	Preservative	(74)
106	Materials (a) Nails	(72)
	(b) Paint	(72)
107		

D—*Sundry*

108	Sundry materials	(52)
109		

110–119 *OVERHEADS*—Woodlands, nursery and sawmill. These can be split into the three groups if required.

Labour additions
- 110 WET TIME
- 111 HOLIDAYS WITH PAY
- 112 ILLNESS PAY
- 113 TRANSPORT OF WORKERS (to and from work)
- 114 PENSIONS PAID TO FORMER EMPLOYEES
- 115 EMPLOYER'S LIABILITY INSURANCE
- 116 EMPLOYER'S CONTRIBUTION TO PENSION FUND OR SUPERANNUATION SCHEME
- 117 PERQUISITES
- 118 BUILDINGS (a) Workers' houses—realistic rental (e.g. not less than Gross Annual Value, less rent, plus rates)

APPENDIX I 177

Forest operation no.

  (b) Sawmill—realistic rental (e.g. not
   less than Gross Annual Value, less
   rent, plus rates)
  (c) Other buildings actual expenditure
   in the year
119

120–129 *Supervision*
- 120 SUPERVISION BY WORKING FOREMAN (20, 21)
  and/or WORKING HEAD FORESTER
- 121 SALARY OR PROPORTION OF WOODLAND
  MANAGER and/or AGENT
- 122 OWNER VALUE OF SERVICES (if any)—
  TIME SPENT ON SUPERVISION AND
  ADMINISTRATION
- 123 DIRECTOR'S SALARY and/or CONSULTANT'S
  MANAGEMENT FEES
- 124 OFFICE STAFF
   (a) Salary
   (b) Perquisites
- 125 STAFF (a) National Insurance ⎫
     (b) Illness pay   ⎬ if charged separately
     (c) Holiday pay   ⎭
- 126 STAFF (a) Employer's liability insurance
     (b) Pension Fund
- 127 HOUSES—Supervisory staff—realistic rental
    (or proportion), less rent, plus rates
- 128 TRAVELLING EXPENSES, including car allowances
   (a) Woodland Manager/Head Forester
   (b) Agent
   (c) Owner
- 129

130–139 *Office and Miscellaneous Expenses*
- 130 OFFICE BUILDING—realistic rental plus rates
- 131 OFFICE MAINTENANCE (a) Light
           (b) Heat
           (c) Water
- 132 COMMUNICATION EQUIPMENT
   (a) Stationery
   (b) Postage
   (c) Telephone(s)
   (d) Typewriter depreciation
   (e) Desk Calculator depreciation
- 133 BOOKS AND TECHNICAL MAGAZINES
- 134 AUDIT AND LEGAL FEES
- 135 SUBSCRIPTIONS TO SOCIETIES (excluding
vermin eradication society—56)

Forest operation no.

136 INSURANCE (a) Office
                  (b) House
                  (c) Sawmill
                  (d) Other building
                  (e) Public liability
137 BANK CHARGES—excluding interest on overdraft
138 MISCELLANEOUS (a) Advertising
                              (b) First aid kit
                              (c) Bad debts
139

140–149 *LAND, tax and valuation*
140 RENT
141 LAND TAX, e.g. for the U.K.
     (a) SCHEDULE "B" TAX (based on prairie value of land)
     (b) SCHEDULE "D" TAX   } U.K. only
     (c) STIPENDS PAID (TITHE REDEMPTION)
         —only on former farm land
142 INTEREST ON OVERDRAFT OR BORROWED MONEY
143 VALUE OF WOODS
144 ANNUAL INCREASE/DECREASE IN THE CAPITAL VALUE OF THE FOREST
145 VALUE OF NURSERY STOCK
146 ANNUAL INCREASE/DECREASE IN THE CAPITAL VALUE OF THE NURSERY
147 VALUE OF THE SAWMILL STOCK
148 ANNUAL INCREASE/DECREASE IN THE CAPITAL VALUE OF THE SAWMILL STOCK
149

INCOME

Sales to outside buyers should be separated from sales to the "Forest Enterprise" and sales to the Estate outside the "Forest Enterprise". Therefore it is proposed that capitals A, B and C should prefix the standard heads:
    A   for sales to outside buyers
    B   for sales to the Forestry Enterprise
    C   for sales to the Estate outside the Forestry Enterprise (sawmill, farm, etc.)

			Units sold
200–209	*Sales of Timber—Woodlands*		
	Unconverted Timber (thinnings and final fellings)		
	200	Standing	Volume
	201	Blown	Volume
	202	At stump	Volume
	203	At Rideside/Roadside	Volume/Weight
	204	Delivered	Volume/Weight
	Converted Timber (woodlands)		
	205	Poles	Volume and number

APPENDIX I

			Units sold
206	Fencing material	(a) Posts (b) Strainers (c) Stakes	Volume and number
207	Decorative material	(a) Rustic fencing (b) Fencing rails (c) Hurdles	Volume and number
208	Commercial wood	(a) Pitprops (b) Pulpwood	Weight
209			

210–219 *Other Sales of Forest Products*

210	Bark	Weight
211	Firewood (sold during scrub clearance)	Area/weight
212	Cut firewood (a) Cordwood or long firewood (b) Firewood blocks	Weight
213	Christmas-trees	Each
214	Greenery and moss	Bundles
215	Cones (and seeds)	Weight
216	Humus	Weight
217	Bean sticks, etc.	Weight
218		
219		

220–229 *Sawmill*

220	Unconverted timber	Volume/weight
221	Fencing material, etc. (a) Posts (b) Strainers (c) Droppers (d) Stakes (e) Poles	Volume/number
222	Sawn timber—through and through	Volume/grade
223	Sawn timber for construction	Volume/grade
224	Sawn timber—sleepers	Each
225	Commercial wood (a) Pitprops (b) Mining timber (c) Pulpwood	 Volume/grade Volume/grade Weight
226	Constructed products (a) Gates (b) Water boxes	Each
227	Firewood (a) Round (b) Sawn (c) Slabs	Weight
228	Waste products for panel products, etc. (a) Sawdust (b) Chips	Weight
229		

180 COST AND FINANCIAL ACCOUNTING IN FORESTRY

				Units sold
230–239	*Nursery*			
	230	Seedlings		Number
	231	Transplants		Number
	232	Cuttings		Number
	233	Shrubs		Each
	234	Plants		Each
	235	Christmas-trees		Each
	236			
	237			
	238			
	239			
240–249	*Stores*			
	240	Bought in timber		Number
	241	Fencing wire	(a) Plain	Rolls (length)
			(b) Barbed	
			(c) Netting	
	242	Metal stores	(a) Staples	Weight
			(b) Butterfly ratchets	Each
			(c) Nails	Weight
			(d) Tying wire	Weight
	243	Road materials, etc.	(a) Culverts	Each
			(b) Clay pipes	Each
			(c) Gravel	Weight
	244	Vermin destroyer	(a) Snares	Each
			(b) Poison	Weight
			(c) Cartridges	Number
	245	Insect and plant destroyers		
		(a) Weed killers		Weight
		(b) Soil sterilisers		
		(c) Sprays		
	246	Fertilisers		Weight
	247	Seeds and cones		Weight
	248	Preservatives, etc.		Weight
	249			
250–259	*Machinery Stores*			
	250	Petrol		
	251	Diesel		
	252	Oil		
	253	Lubricants		
	254	Grease		
	255	Feed		
	256	Spare parts		
	257	Tyres		
	258	Hand tools, etc.		
	259			

APPENDIX I

Units sold

260–269 *Grants and Miscellaneous Receipts*
- 260 Planting Grant ⎫
- 261 Management Grant ⎬ United Kingdom only
- 262 Small Woodlands ⎭
- 263 Other grants
- 264 Rents (a) House
     (b) Sporting
     (c) Grazing
- 265 Sickness and other benefits
- 266 Hiring out of labour to other departments
- 267 Hiring out of vehicles and/or machines
- 268 Other receipts, e.g. sale of vermin
- 269 Tax repayments

Appendix II

Capital Valuation Comparison Between the Actual Value and Potential Value Methods of a Sitka Spruce Plantation Yield Class 18 m³/ha

In this capital valuation comparison (Table 36) the costs and revenues have been taken from the main text, Tables 15b and 27. The plantation can either be looked on as a 1-hectare spruce plantation proceeding through its lifetime or as a "normal" Sitka spruce forest with an equal representation of 1 hectare in each age class. The rotation age of maximum financial yield is 34 years and the financial yield is 10.415%. This rotation age does not coincide with the rotation age of maximum mean annual increment—18 m³/ha—which is 55 years.

The actual capital value in Table 36 is extracted from Tables 15b and 27 whereas the potential capital value has been arrived at by increasing the net capital investment by 10.415% per year. The positive value of £1 at year 34 indicates that the maximum financial yield interest rate is marginally in excess of 10.415%. If the rotation age had been fixed at 35 years then the financial yield would be about 10.32% and the income from the final felling £4257.

It has been assumed that after year 16 the capital investments for direct and overhead operations are financed from the income from thinnings.[1]

It will be noticed from Table 36 that the potential capital value is substantially greater than the realisation or actual capital values, except in year 0 and just before final felling. This is illustrated diagrammatically

---

[1] Strictly speaking there should be costs and therefore net capital investment in the years when no thinnings take place. If we assume that the costs amount to £1.33 per year from year 17 to 34, and that the gross income is £44 year 17, £54 year 20, £110 year 23, £231 year 27, £304 year 31, and £4008 year 34, then the financial yield will be marginally less than stated above.

TABLE 36. *Capital Value of 1-hectare Sitka Spruce Plantation Y.C. 18 m³/ha Units £. Constant money values*

Age yrs	Net capital investment (N.C.I.)	Capital valuation (end year)		Age yrs	Net capital investment (N.C.I.)	Capital valuation (end year)		Net income
		Actual value (A.V.)	Potential value (P.V.)			Actual value and realisation value (A.V. & R.V.)	Potential value (P.V.)	
0	170.16	170	170	17	−40.00	300	997	40
1	1.33	175	189	18		357	1101	
2	7.08	185	216	19		416	1216	
3	1.33	190	240	20	−50.00	580	1293	50
4	1.33	195	266	21		669	1427	
5	1.33	200	295	22		757	1576	
6	1.33	205	327	23	−105.00	975	1635	105
7	1.33	210	363	24		1106	1805	
8	1.33	216	402	25		1238	1993	
9	5.00	225	449	26		1729	2201	
10	1.33	230	497	27	−225.00	1658	2205	225
11	1.33	236	550	28		1822	2435	
12	1.33	242	608	29		2501	2688	
13	1.33	248	673	30		2743	2968	
14	1.33	254	744	31	−300.00	2635	2977	300
15	1.33	260	823	32		2852	3230	
16	30.50	295	940	33		3069	3630	
				34	−4007.00	0	1	4007

in Fig. 11. In practice the value of a forest should lie somewhere between the potential and actual values.

When evaluating profit for any one year the manager is interested in the change in capital value from one year to the next and this should not be greatly different whichever method is chosen although the balance-sheet will vary depending on the method of valuation used. However, once a method has been chosen it should not be changed to suit different financial statements or different years' results.

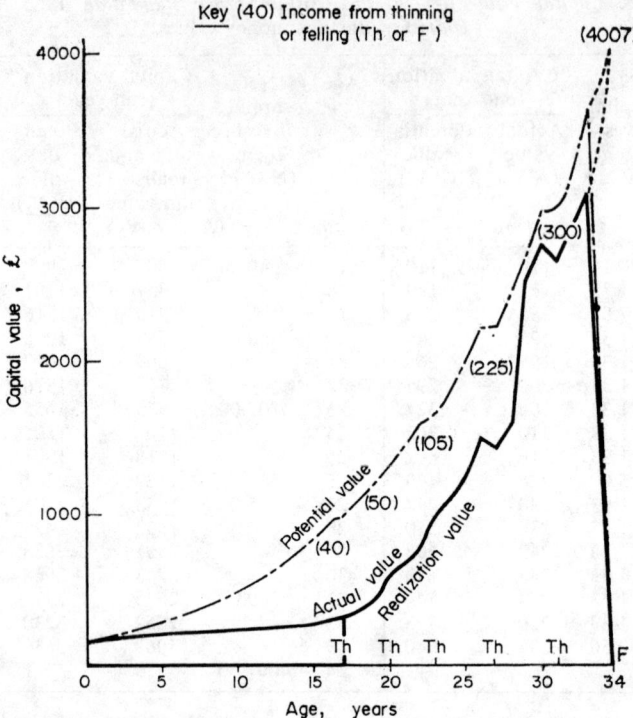

Fig. 11. Capital valuation comparison between potential, actual and realisation-value methods. (Sitka spruce plantation of 1 hectare valued each year throughout its lifetime of 34 years. Constant money values assumed for potential-value method.)

INDEX

Account
   capital 85, 122
   external 79–91
   income and expenditure xv, 79–91, 121
   internal 39–53
   machine 84
   profit and loss xv, 79, 91, 121–122, 127
   trading 79–91
Accountant's capital value 95, 120
Actual capital value 94, 95–111, 182–184
Administrative expenses 82, 88
Age, definition and assumption 128, 129
Allowance for crop damage *see* Volume loss
Analysis
   compartment 33
   cost/benefit xvi
   costing xv, 3–78
   financial xv, 79–126
   marginal xvi, 74
   market xvi, 74, 75, 165
   of primary records 31–38
   operation 33–34, 36, 37
   sector xvi, 39–53
   summary 34
   wage 33, 34
Appropriate rate of interest 101, 103, 104, 107
Area allocated overheads 54, 63–65, 66, 67, 68
Average costs *see* Cost

Balance sheet xv, 122, 123
Barking 4, 8, 172
Beating up 5, 6, 14, 33, 35, 36, 37, 48–51, 157, 171

Bonus rates 62, 71, 72
Bradley, R. T. 98, 99, 150, 164
Brashing 4, 6, 14, 36, 37, 41, 157, 171
Budget programming xvi, 73

Capital
   account *see* Account
   cost xv, 130, 136, 155, 158
   equipment 18
   expenditure *see* Capital cost
   expenditure—depreciation *see* Depreciation
   useful life 19
   valuation xv, 79, 91, 92–120, 182–184
Capitalisation value 61, 95, 119
Cash book xv, 79
Cleaning (early and late) 5, 6, 171
Compartment
   analysis sheet *see* Analysis
   costing schedule 42–43
   costs xv, 39–53, 63–65, 99, 100, 106, 108
   forms *see* Compartment costs
   records xvi, 46, 48, 49, 50, 51, 70–99
   summary sheet 33
Compound
   interest 127, 130
   process 128, 158
Constant
   costs *see* Cost
   prices (revenue) 137, 144, 147, 148
Consumable stores 82, 88
Cost
   analysis *see* Analysis
   assumptions 129, 130
   average 47, 75, 129, 130
   /benefit analysis *see* Analysis
   comparison xvi, 71

INDEX

Cost—*contd.*
  components 7
  constant 102, 117, 118, 129–130, 132, 135, 136, 144, 147, 158, 184
  contract xv, 17, 33, 69, 80
  current 105, 132
  direct xv, 12, 39–53, 68, 80, 83, 170–181
  establishment 5, 49, 51, 98, 168, 170, 171
  harvesting 172
  historic 100, 102, 105
  initial establishment 48, 50, 157
  labour 32, 33, 35, 36, 56, 58, 62
  machine 84
  marginal 75
  nursery 173, 174
  operation xv, 47
  overhead *see* Overhead cost
  planting *included under* establishment cost
  reduction 164, 165, 167
  running xv, 80
  sawmill 174
  tending 6, 171
  unit 43, 44, 46, 48, 49, 50, 51
  variations 148, 149
  vehicle 81, 174–175
Costing
  schedule xv, 40, 42, 43, 52, 53, 64–66
  unit 5, 6, 42, 44, 46, 48–51, 171–180
Current
  annual increment 144, 145
  prices 137

Depreciation 18–30, 60, 61, 84, 87
  capital expenditure 81, 86, 87
  factors 19
  methods
    annuity 25
    declining value 21
    production 23–24
    reducing balance 21
    straight line 20
    sum of digits 23
Direct
  cost *see* Cost
  overhead cost *see* Overhead

Discount
  process 127, 128, 149
  rate 75
Discounted
  expenditure 127–136, 143–147
  felling value 139, 140
  income 137–142, 143, 144–147
  thinning value 139, 140

Establishment cost *see* Cost
Expectation value 94, 116–117
External account *see* Account

Factors of production xv
Figures ix
Final felling 137–142, 162, 182
Financial
  planning xvi
  yield xv, 70, 75, 127, 143–153, 154–158, 160–163, 169
Food and Agricultural Organisation of the U.N. F.A.O. 165
Forecasting 165
Formulae xii

Grants xv, 88, 89
Grayson, A. J. 98, 99, 150, 164

Hart, C. E. 164
Harvesting cost *see* Cost
Headquarter overhead *see* Overhead
Historic cost *see* Cost

Immature phase (plantation) 99–108, 110
Income *see* Revenue
  elasticity 152
  and expenditure account *see* Account
Infinite rotations 133, 145, 146
Infinity formula *see* Formulae
Inflation 28, 99, 101, 102, 103, 104, 129, 137, 149–152, 155, 158
  multiplier 100, 102
Initial establishment cost *see* Cost

INDEX

Internal
  (cost) account *see* Account
  rate of interest (owner's rate) 160, 161
  rate of return *see* Financial yield

Johnson, D. R. 98, 99, 150, 164
Journal xv, 79
Labour
  additions 56, 60
  cost allocated overheads *see* Overheads
Land valuation 108, 109
Ledger xv, 79
Leslie, A. J. 61

Machine
  account *see* Account
  book xv, 15, 33
  cost *see* Cost
  input 74
Management tables 139
Manpower planning 74
Market
  analysis *see* Analysis
  intelligence 165
Material
  book xv, 16, 33
  input 74
  planning 74
Maximum financial yield *see* Optimum rotation
Mean annual increment 104, 144, 145, 182
Merchantable
  crop valuation 96–98
  phase 100, 102, 104, 110
Multiplier factors 131, 132, 134, 140

Natural forest valuation 109
Net
  capital cost 100
  discount revenue (N.D.R.) 138, 159–164
  present worth *see* N.D.R.
Normal forest 90, 114, 119, 123, 127, 182

Nursery
  cost *see* Cost
  costing schedule *see* Costing schedule

Operation
  analysis sheet *see* Analysis
  cost *see* Cost
  costing schedule *see* Costing schedule
  summary sheet 33
Operations 3–9, 170–181
Optimum rotation (maximum financial yield) 142, 143, 153, 158
Overhead
  allocation form 55, 56
  allocation proportional to labour costs 54, 62, 63, 64, 65, 66
  allocation proportional to revenue 54
  allocation proportional to volume 54
  area allocation 54, 55, 63, 64, 65, 66, 67, 68, 130, 131
  cost xv, 9, 12, 54–69, 80, 83, 149, 170–181
  direct cost allocation 54, 62, 63, 64, 65, 66
  district 54, 67
  headquarter 54, 67
  interpretation 68
  organisational level 54
  origin 54
  project 54, 67
  regional 54, 67
  research 54, 67

Patterson, A. R. 15
Piece work 62, 71, 72
Planning *see* Machine, Manpower *and* Material planning
Planting
  costs *see* Costs
  distance 166, 167
Potential
  capital value 94, 102, 112–115, 182, 184
  return 123

Price
  assumption 137
  /size gradient 105, 137, 143, 145, 158, 168, 169
Primary records *see* Records
Productivity measurements xvi, 72, 73, 129, 136, 152, 167
Profit and loss account *see* Account
Profitability degree 90
Project
  overhead *see* Overhead
  planning xvi
Proportional overhead *see* Overhead

Real return 150, 152
Realisation value 94, 111, 182–184
Records, primary xv, 10–17, 31–38
Regional overhead *see* Overhead
Rent xv, 57, 59, 60, 61, 82, 88
Research overhead *see* Overhead
Returns xv, 144, 147
Revenue
  allocated overhead *see* Overhead
  expenditure 120
  (income) 88, 139, 149, 151, 157, 164
Rotation 103, 122, 128, 132, 133, 135, 140, 143, 144, 147, 149, 152, 155, 156, 164, 182
  infinite 133
  of maximum financial yield 128, 143–153
  single 132, 133

Sale of timber 89
Sawmill
  costing schedule *see* Costing schedule
  costs *see* Cost
Sector analysis *see* Analysis
Standard heads 3–9, 170–181
Stumpage
  price xvi, 71
  rates 75
Summary analysis sheet *see* Analysis sheet
Supervision 56, 57, 60
Surplus/deficit 90

Tables x–xi
Taxes 58, 60, 61, 156
Tax rebates xv, 88, 89
Tending costs *see* Cost
Thinning 137–142, 156, 162, 169
Timber
  prices 150, 152
  sales 88
Time
  sheets xv, 4, 11, 12, 13, 14, 33, 56, 74
  unit 11, 12
Tools 57, 59, 60
Trading account *see* Account

Uneven aged stand valuation *see* Valuation
Unit
  area capital valuation 101, 106, 110
  cost *see* Cost
  overhead *see* Area allocated overheads
Useful life of capital 19

Valuation
  capital *see* Capital valuation
  closing 109–110, 122
  increase/decrease 110, 121
  of the land 108, 109
  of natural forests 109
  of uneven aged stands 109
  of the whole crop 109–110
  opening 109–110, 122
Vehicle cost *see* Cost
Volume
  allocated overhead *see* Overhead
  loss (crop damage allowance) 138, 154
  yield (Sitka spruce) 138

Wage
  analysis sheet *see* Analysis
  summary sheet 33, 34, 35
Wardle, P. A. 166
Whole crop valuation *see* Valuation
Woodland costing schedule *see* Costing schedule

Yield table 139
Young woodland valuation 98